THE DICTIONARY OF DANGEROUS IDEAS

BY MIKE WALSH

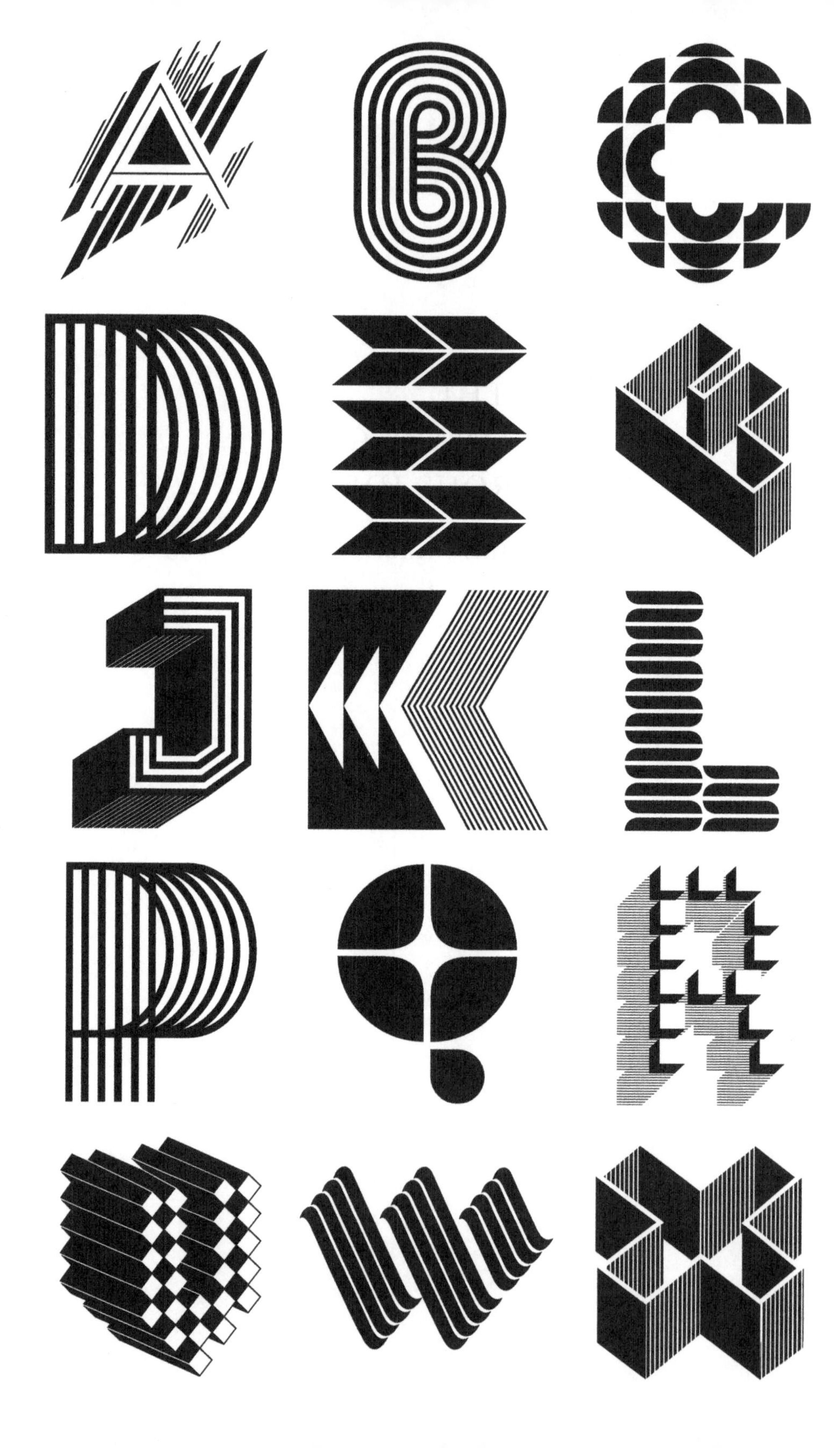

The Dictionary of Dangerous Ideas
Mike Walsh

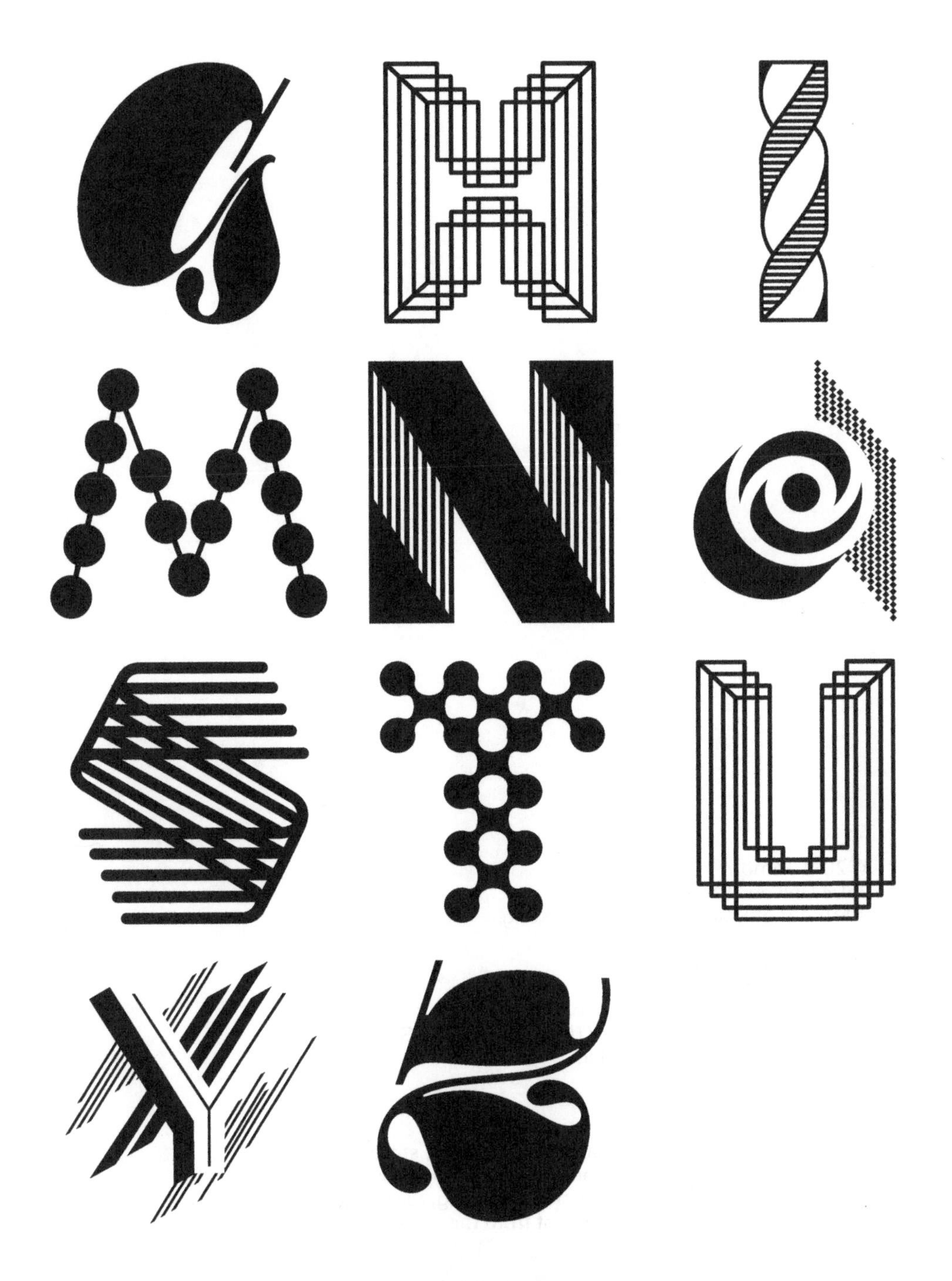

2504 Universal Trade Centre,
3 Arbuthnot Road, Central, Hong Kong.

Photos and text by Mike Walsh.

Design by Bold (boldstockholm.se).
Alphabet by Gemma O'Brien (The Jacky Winter Group).
Editing by Lucy Howard-Taylor.

A/B Test
Adaptive Learning
Advanced Manufacturing
Algorithm
App
Artificial Intelligence
Automation

Big Data
Biohacking
Biomechatronics
Blockchain
Botnet

Carbon Nanotube Computer
Citizen Science
Clickbait
Cloud
Co-creation
Cohort Effect
Collaboration Map
Computational Design
Computational Thinking
Consumerization
Contextual Computing
Crowdfunding
Crowdsourcing
Cyptocurrency

Data Profiling
Deep Web
DevOps
Digital Uprising
Divergence
Drone

Emoji
Enterprise Social Network
Enterprise Virtualization
Exascale
Experience Design

Fab Theft
Freelancers

Gamification
Genome Editing
Growth Hacking

Hacktivist
High Frequency Trading
Holacracy

Inbound Marketing
Innovation Ecosystem
Internet of Things

Junk Patent

K-Scale

Life Extension

Maker Culture
Mesh Network
Metadata
Microsat
Mobile First

Nanofactory
Net Neutrality

Open Platform

Permissionless Innovation
Personal Robotics
Pharmacogenomics
Pivot
Predictive Analytics
Programmatic Marketing

Quantified Self
Quantum Computer
Qwerty Effect

Radical Transparency
Remote Work
Re-shoring
Retro Modding

Self-Driving Car
Shanzhai
Sharing Economy
Smart Tattoo
Space Commercialization
Synthetic Neocortext

Talent Density

Ultra Wealth
Urban Scaling
User Agreement

Virtual Reality

Webscale
Workforce Analytics

XAAS

Yottabyte

Zero Email

Mike Walsh

MIKE WALSH IS the CEO of Tomorrow, a global consultancy on designing business for the 21st century. He advises leaders on how to thrive in this era of disruptive technological change. Mike's clients include many of the Fortune Global 500, and as a sought-after keynote speaker he regularly shares the stage with world leaders and business icons alike. Mike is a board member and strategic investor in the North Alliance, Scandinavia's leading digital marketing group. All the photos in this book were taken on his travels with a beaten but beloved Leica.

WWW.MIKE-WALSH.COM

Introduction

WHAT MAKES AN idea dangerous is often not so much the idea itself as the troubling questions that it raises. Questions about the nature of reality, authority, society, and the boundaries between truth and heresy.

History is full of political heroes, social revolutionaries and religious martyrs who have challenged those in power – with their ideas, actions and in many cases, their very existence.

And yet, some of the greatest heretics were actually scientists. Science may be the realm of the objective and the verifiable - but sometimes, the more self-evident an idea appears in retrospect, the more dangerous it was when first proposed.

Consider astronomy. Once telescopes were invented, it was inevitable that someone would notice that the earth revolved around the sun and not the other way around. And yet what made that idea so dangerous to think, let alone publish in the early 17th century, was not the principle itself as its conflict with established religious dogma. To expound his theories, Galileo Galilei wrote his book as a dialogue between a character who shared his ideas, and another named Simplicio (simpleton in Italian), who represented the Church's position. At that point, Galileo's fate at the hands of the Inquisition was as inevitable as his dangerous line of thought.

Arriving from Europe as an immigrant, Nikola Tesla brought with him an idea for a newer and more efficient form of power generation known as alternating current. His radical thinking set him against Thomas Edison, who had developed a competing technology of direct current. Edison decided to ruin Tesla – at one point staging a national tour of animals tortured live on stage with his rival's technology.

It was not the last time Tesla's dangerous ideas would provoke the ire of the rich and powerful. When JP Morgan, who owned a vast portfolio of copper mines, realized that Tesla's plans for wireless free energy distribution would not only threaten his opportunity to profit from copper transmission lines but the fortunes of electricity companies – he persuaded Wall Street to turn their backs on him.

Ideas can also be considered dangerous when we don't or refuse to understand the people thinking them.

Alan Turing, the founder of modern day computing, was given the diabolical choice of staying in prison or being chemically castrated after being convicted of homosexuality in the 1950s. Even when released, his revoked security clearance made it impossible for him to continue his work on cryptography. He committed suicide the year after.

There have always been dangerous ideas and dangerous thinkers – but the reason I wrote this book is because I believe we are standing at a point in time when the pace of technological innovation will be matched by the growing conflict that radical ideas will create in our world.

If there is one idea that links the ones that follow, it is this: everything is now capable of being expressed digitally. This means that anything can be defined by software, manipulated by algorithms, and cheaply analyzed by massive computational resources.

We no longer need to wonder about the position of the planets or the origins of life. Instead we can re-program nature, re-configure our bodies and even design synthetic life forms and intelligences. When reality itself can be re-written like software, the results will be hard to distinguish from magic.

As we watch these events and discoveries unfold, we will be excited and then fearful. We will seek to make both heroes and villains out of those that would refactor the world, and in doing so, create both billionaires and martyrs.

And yet, unlike in previous centuries, attempts at redaction or censorship will be an exercise in futility. Dangerous thinking has always been open to anyone, but the tools of transformation are now in the hands of everyone.

Suppressing ideas has never been a surety against a future we fear. If anything, the opposite is true. The future thrives on ideas that challenge it.

Oscar Wilde said it best - an idea that is not dangerous is unworthy of being called an idea at all.

A/B Test
Adaptive Learning
Advanced Manufacturing
Algorithm
App
Artificial Intelligence
Automation

A/B Test

PEPSI-COLA
MADE WITH Real SUGAR
Diet Coke
20 FL OZ (1.25 PT) 591 mL

pepsi
pepsi
pepsi
Diet Coke
Diet Coke
20 FL OZ (1.25 PT) 591 mL

An A/B test is a random experiment in which two alternatives are tested to see which performs better.

WHAT IF YOUR best strategy were leaving things to chance?

After giving a talk in Oslo, I shared the back of a taxi with Jim Messina – who had been in charge of Obama's presidential re-election campaign. As we rounded the streets of the Norwegian capital, Jim explained they had met with a wide range of gurus including Steve Jobs, Steven Spielberg and Eric Schmidt. Spielberg had this advice: voting for Obama the first time was like touring with the Rolling Stones in the 60s. But the second time around things would be tougher. The rockers were more famous, but also older. And worse, the concert tickets were now too expensive. In other words, to engage the nation - they had to make the campaign sexy again.

One of the ways they decided to do this was with data. In the first election, they had six data analysts. This time around, Jim hired eighty. Every time they sent out an email for funding or support, they would try multiple versions of subject lines, wording formats and calls to action. They would A/B test the variations, and adjust accordingly.

A/B or "split testing" is nothing new. The first A/B test was run at Google on February 27, 2000 – and was a disaster. Google wanted to know whether showing fewer results on their search page would lead to better engagement. So a select percentage of visitors were shown just 20 results, but due to a technical error, the pages loaded more slowly for the experimental group – leading to the unexpected insight that even a tenth of a second difference in response time could significantly impact someone's perceived experience.

If you have ever had an argument with someone about what your customers really want, an A/B test is a good way to even the score. But be warned. In an increasingly data-driven society, you can never be sure that you yourself are not part of someone else's grand experiment. Did you receive a slightly different email this morning from your bank? Did your Amazon homepage offer you fewer purchase options than a friend's? Or, as some Facebook users discovered recently – did your timeline deliberately show you negative stories to see how you responded emotionally?

If you were to A/B test your most firmly held view about what your customers wanted, what would be the result?

Adaptive Learning

Adaptive learning is a data-driven approach to education that personalizes teaching based on the real time progress and capabilities of a student.

WHAT IF EVERYONE'S education were personalized?

I was lucky to have some wonderful teachers growing up. My English teacher took me to the theatre so I could understand what Shakespeare's words on the page were really for. I had an ancient history scholar who painstakingly wrote me notes on Cicero's art of persuasion. And an economics expert, who not only explained the mysteries of money but also taught me mnemonic memory tricks so I could pass my exams with ease.

When I say I was lucky, I really mean it – because none of these teachers were actually at my school. They were employed by my mother, who being Chinese, believed strongly in over-education. Up until that point, I had been an average student. Personalized instruction changed everything. I was not studying. I was learning. And I loved it.

Learning is a very personal activity. Humans do not ingest ideas by simply being exposed to them. They need to integrate new concepts into existing ones, and relate novel material to established neural pathways. That poses a problem for any educational system, because it is not easy or cost effective to scale individual teaching.

Adaptive learning is one potential approach. By using data, analytics, and an understanding of behavioral science - adaptive curriculums re-organize traditional learning by delivering lessons based on what the student needs to learn next in order to make progress. The approach can be non-linear. It may be that to improve your mathematics, what you need is not more algebra but some instruction in linguistics or logic. Or, instead of always learning new things, coaching in memory management so you can better recall the things you have already learned.

Adaptive learning addresses one of the biggest paradoxes that educators face today. Never before in history has so much information, knowledge and learning been so readily and freely available online. And yet – how do you organize all of this into a logical learning plan? And even more importantly, into a curriculum tailored for an individual?

If you were running a charity focused on improving education in the developing world, how would new digital technologies change your approach?

Advanced Manufacturing

Advanced manufacturing utilizes computation, automation and materials science to reimagine the design and assembly of products.

WHAT IF FACTORIES were more like laboratories?

The idea of the factory has a complex past. On one hand, factories symbolize progress, economic growth, jobs and productivity. On the other, they also conjure images of the "dark Satanic Mills" of the Industrial Revolution, child labor, toxic pollution and more recently – worker suicides at Foxconn's facilities that build the world's iPhones.

Factories in the 21st century are rapidly changing, because manufacturing itself is undergoing a technology-led revolution.

Take a tour of Elon Musk's Tesla Model S factory and you will witness a gleaming army of robotic workers and advanced workstations. Massive rolls of aluminum enter at one end, and less than a week later, fully finished cars exit from the other. What is extraordinary is that almost all the parts of the car are made from raw materials on site. Production is agile and re-configurable. The same robot that places a seat in the car is able to adjust and immediately fit a windshield as well.

The catalyst for advanced manufacturing has been the rapidly falling cost of computation, making automation, computer control and digital design affordable options for many industries.

Whether it's for food processing or aerospace engineering, mobile phones or medical devices – the same technologies and techniques you might have once expected to find in a billion dollar chip fabrication lab, are now appearing in conventional factory environments.

How will the rise of advanced manufacturing change the economics and the location of the world's production centers?

Algorithm

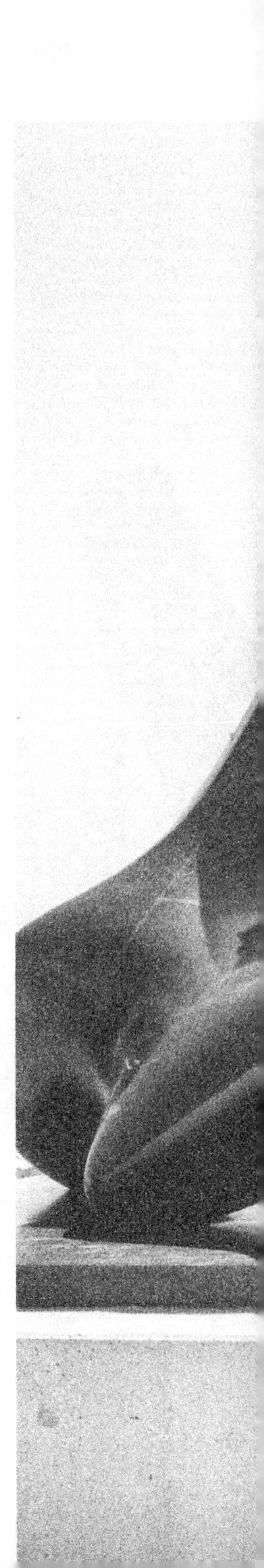

An algorithm is a precise set of instructions that enables a computer to solve a problem.

WHAT IF COMPANIES competed in the future on their ability to design great algorithms?

Algorithms surround us. They determine the search results that appear when we look for something on Google, the books and movies Amazon or Netflix recommend to us, the content that appears on our Facebook newsfeeds and the types of advertisements we are shown.

Ubiquitous and unseen, algorithms are the thought forms of the digital world. The word comes from "algoritmi", the Latin form of al-Khwārizmī, the Arabic mathematician who invented algebra and was among the first to use the concept of zero.

Algorithms exist because computers require very precise instructions in order to work properly. However, they are not just recipes. Algorithms represent the deep insights that give life to modern software.

Consider some of these challenging questions. How do you work out what content on the Web is worthwhile? The founders of Google had the insight to create an algorithm that adapted the peer review model of citations, in order to rank pages by their number of Web links. How do you secure confidential communications? Phil Zimmermann, the creator of the email encryption software Pretty Good Privacy (PGP) created a cryptographic algorithm based on the metaphor of a public-key exchange.

Often the most important difference between technology companies is the unique concept behind their algorithms.

Whether it be placing advertisements, predicting customer behavior, anticipating equipment failure or even pricing products and services, the most useful people in your strategy team in the future might not be the MBA graduates, but those with the degrees in pure mathematics.

What kinds of problems in your business might be better solved by programmers and algorithms rather than management consultants and slide presentations?

App

NO CYCLE
MOBILE PHONE STRICTLY
PROHIBITED
IN

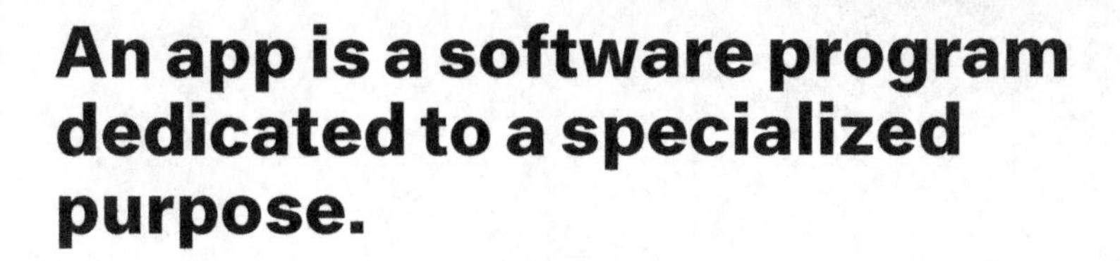

An app is a software program dedicated to a specialized purpose.

WHAT IF THE less something did, the more useful it was?

Software started out simply enough. A video game was two moving lines with a ball in between. A social network was a bulletin board. A word processor was a command prompt rather than a typewriter. And you could choose whatever color you wanted your retro screen text to be – as long as it was green or amber.

With faster processors and more memory, software developers rapidly expanded their feature sets. The result was sprawling cathedrals of code that few used, let alone really understood. Curiously, mobile devices have helped to turn the clock back. With small screens, limited download speeds, and a limit on what users will pay for mobile content, developers have been forced to embrace a culture of applications, or apps.

The app challenges a fundamental tenant of software, namely that the more features and functions you have at your command, the better a program is. Apps do the opposite. They eliminate all extraneous purposes and focus on a single activity or underlying behavior.

The behavioral piece is important to understand. One of the reasons Facebook spent so much money buying applications like Instagram and Whatsapp, is that they realized their biggest threat was becoming too big to see the next new trend. Brands are not viral, new behaviors are – whether it is sharing messages that delete themselves, or retro-fitting your photos with filters.

When you have hundreds of millions of users, making interface changes is a political and controversial process, equivalent – in the words of engineers at Facebook – to re-arranging millions of people's living room furniture at once. Apps provide a more nimble path to experimentation and viral adoption – not just for social networking giants, but for all kinds of companies as well.

If you had to simplify your company's core offering to just an app, what would it do?

Artificial Intelligence

www.escueladesirenas.com.ar

Artificial intelligence is the capability of a machine to emulate human qualities such as perception, pattern recognition and judgment.

WHAT IF MACHINES could make better decisions than us?

You might remember Eliza. It was a clever conversation program based on the conversational style of a clinical psychologist. At first blush, a solid step to creating true machine intelligence. But as it turns out, getting computers to do human parlor tricks is relatively easy. What is much harder is developing the computational capacity for every day tasks that we take for granted, like recognizing images or interpreting human speech.

That's why most AI research today is focused on learning. Previously, scientists wrote very complicated programs to help computers perform individual tasks. That took a lot of time, and also left computers unable to deal with more ambiguous data.

One of the more interesting theories at present is that the human brain does not have thousands of programs for thousands of different things, but one influential and complex learning algorithm that powers perception.

Neural networks were a popular idea in the 1970s, but it has only been with the computer power and programming developments of late that researchers have been able to create "deep learning" models that make better use of layers of virtual software neurons.

You can teach a neural network in two ways. One is using labeled data. For example Andrew Ng, who founded Google's digital brain team – ran an experiment where he took about 50,000 pictures of coffee mugs in the Bay Area and fed it to his computer in order to build a coffee mug detector. The computer became very good at spotting mugs, but it is hardly a scalable approach for the real world. After all, most humans or animals don't have to first see thousands of versions of something before they know what it is.

So Ng's team tried the second way of teaching a neural network. They took a simulated brain, and had it watch YouTube for a week. When they probed the network to see what it had learned, they discovered that the AI had learned what a cat was, even though no one had ever told it what a cat was.

In what circumstances would an artificial intelligence make a better board member than a human being?

Automation

福 茗茶
天福
TEA
TEA

Automation is the use of technology to make a process or an activity run without human intervention.

WHAT IF YOU could replace repetitive human tasks with machines?

Automation takes many forms. Multi-purpose industrial robots on assembly lines; cloud-based HR software that eliminates routine administration; marketing automation platforms that do a better job of warming up sales leads than a three-martini lunch.

The age-old problem with automation is that it often means eliminating jobs. As Henry Ford lamented, "Why is it when I hire a pair of hands, I get a human being as well?" Automation in the 21st century introduces a more subtle complication. Would you keep your job if it meant working more like a machine?

I spoke recently to a group of manufacturers from India and Bangladesh. They were struggling with the idea of automation. They knew on one hand that they had a low cost labor advantage in their domestic market, and yet they feared that this mindset was holding them back from the benefits of new technology. Their solution was partial automation – smarter machinery that minimized the potential of human co-workers to make mistakes.

Even large technology companies have implemented similar systems. Amazon may be a cutting edge, data-driven business but for the people working in the e-commerce giant's warehouses, endlessly scanning boxes and following strict workflows, automation has not replaced human beings – just eliminated the potential for human judgment.

If the true impact of automation is replacing supervision, it is worth asking whose jobs are really at risk. Think about how many layers of middle management exist inside modern corporations, how many endless meetings and pointless administrations. What might all of those people do when the transactional parts of their jobs largely disappear?

If technology could automate most of your job, what would you need to do in order to remain useful?

Big Data
Biohacking
Biomechatronics
Blockchain
Botnet

Big Data

LC TAMBAHAN #1
LC 005A
LC 005B
LC 037A
LC 025A
LC 025B
LC 025C
LC 017A / 1
LC 017A / 2
LC TAMBAHAN #2
LC 017A / 3
LC 050A
LC 026A / 1
LC 026A / 2
LC 026A / 3
LC 019A
LC 001A
LC 001B / 1
LC 001B / 2
LC 001B / 3
LC TAMBAHAN #3
LC 067A
LC 067B
LC 069A
LC 006A
LC 006B
LC 009A
LC 067C
LC 069B
LC IXP
LC SPACE / MATRIX II
LC SP10 #6
LC SP10 #5
INFOTEL B
INFOTEL C

Big Data is a set of data sufficiently large that it defies conventional approaches to management and analysis.

WHAT IF DATA could replace theoretical models?

Data was once scarce. A few hundred years ago scientists were desperate to obtain more of it to understand how the world really worked. Captain James Cook, for instance, was tasked in the 18th century to voyage to the other side of the world – not just to discover new continents, but also to obtain better observational data, so astronomers could calculate the distance of the earth from the sun.

Now we have the opposite problem. Excess data is a byproduct of the modern world. As the price of storage and computation has fallen, everything has become digitized. That has led to data sets so large and complex that it can be difficult to study them. Web giants like Yahoo and Google have been forced to develop new tools to sort and order the data their users are creating, and in doing so, they have noticed something very interesting.

In 2009, Google researchers Alon Halevy, Peter Norvig, and Fernando Pereira wrote an influential paper titled, "The Unreasonable Effectiveness of Data", in which they argued that with enough data, the choice of mathematical model was no longer so important. In the case of automated language translation, for example, they discovered that "simple models and a lot of data trump more elaborate models based on less data."

Chris Anderson, former editor of Wired magazine, went even further, arguing that massive amounts of data and applied mathematics meant that the traditional scientific method of hypothesis, model and test was becoming obsolete. That's a dangerous line of thought because it suggests that as long as you have correlation, you no longer need to worry about causation.

Whatever your view on the debate, the real lesson is that there is power in building applications and services that scale with data. To write a good product recommendation engine, you don't need to have tried every item in the supermarket. But it does help if you understand something about human nature and have a theory about how people with similar tastes might behave.

In your business, what data metric would transform the way you make decisions?

Biohacking

Biohacking is the manipulation of biological systems as if they were software.

WHAT IF WE could re-program nature?

Consider this. The cost of sequencing DNA has fallen from about $100,000 for a million base pairs of DNA code in 2001, to around 10 cents today. What used to require highly specialized equipment can now be done with ready-made kits in less than three days, while all the enzymes and chemicals you need to create, manipulate and stick together chains of DNA can be bought at retail.

Welcome to the world of biohacking – the strange parallel universe where cutting edge science has become a hobby for amateur enthusiasts and a terrifying scenario for authorities who are faced with potential acts of terrorism cooked up in homemade labs.

If all that sounds like the early days of the computer revolution, there are many similarities. When you examine the structure and sequences of DNA it starts to look a lot like digital code - code that you can revise and modify. In a way, genetic engineering is a form of software development.

DNA can now be printed to order. Biohackers use a CAD-like tool to design DNA for a specific purpose, potentially copying a sequence from nature. You can place an order online and within 72 hours, receive a synthetic DNA sample that you can use to transform your bacteria or yeast.

And just like software, there is now an open sourced catalogue with ready-made sequences called bio-bricks. You can string these bio-bricks together and put them into e-coli as a host mechanism. With these tools, bio-hackers can build entire systems of biological devices which turn off or on in reaction to light, produce vaccines and medicine, or even convert algae to fuel.

Ultimately the aim of biohacking is synthetic biology - rewriting the DNA of microorganisms as if it were simply software.

If genomics is to biology what programming is to computers, then what might it take to become the Google of digital genetics?

Biomechatronics

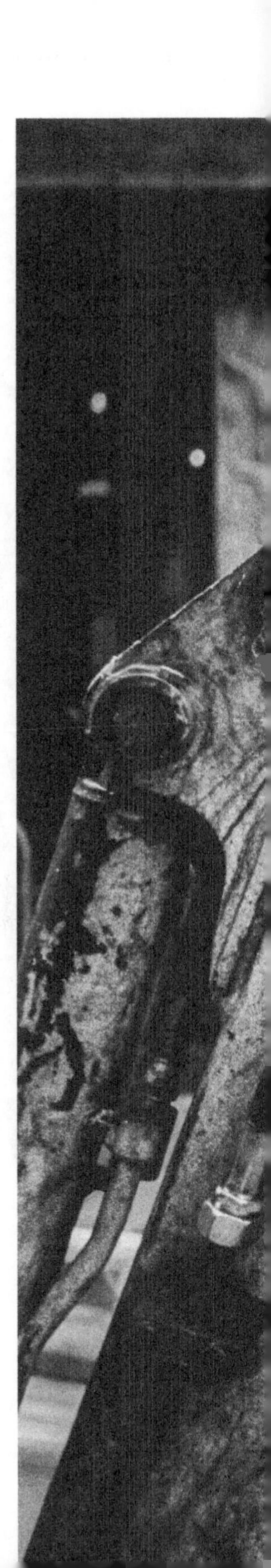

3339

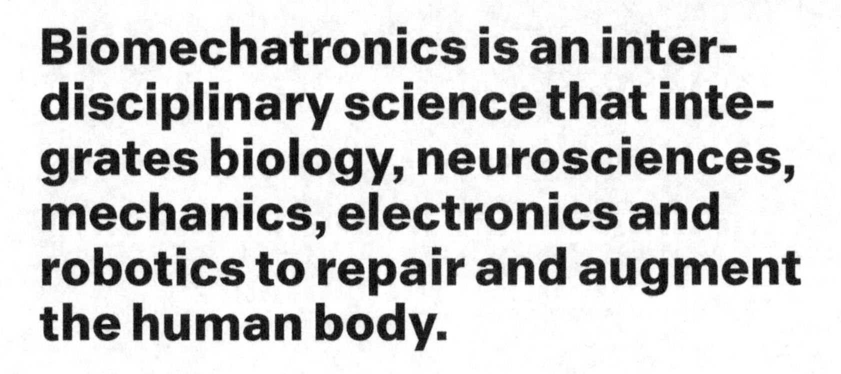

Biomechatronics is an interdisciplinary science that integrates biology, neurosciences, mechanics, electronics and robotics to repair and augment the human body.

WHAT IF ROBOTS could not only replace our lost capabilities, but make us better than we were before?

Hugh Herr lost both his legs to severe frostbite after a rock climbing accident. Rather than be daunted, he realized that the artificial part of his body had in fact become malleable. With his loss, he had gained the ability to create a range of prostheses that were adjustable and would allow him to go beyond the range of standard human capability.

Herr is currently the head of the Biomechatronics lab at the MIT Media Lab and is at the forefront of research into the intersection of biomechanics and neuro control. His devices have helped amputated patients walk with a natural gait, bombing victims dance again, and professional disabled athletes compete against able-bodied peers.

The field of Biomechatronics is a fascinating one to watch as it not only raises complex issues about restoring capability to the disabled, but also the ethics of human augmentation.

Exoskeletons will allow human bodies to be faster, stronger and more capable of feats of endurance. Sophisticated neuro controls will allow us to not only control these robots with our minds, but to eventually be able to develop a sense of touch from our augmented limbs.

For now, the focus of research is figuring out how to embed more of our humanity into bionics – teaching robot limbs to move and respond as naturally as possible. But in the future, the real question may be the opposite. How much of the bionic should we embed into humanity?

Would you accept a world in which soldiers or extreme sports enthusiasts amputated healthy limbs to gain bio-mechanical advantages?

Blockchain

The blockchain is the public record of all the Bitcoin transactions that have ever been executed.

WHAT IF WE no longer needed intermediaries?

In the last few years we have seen the rise and fall of Bitcoin, the dramatic FBI takedown of the Silk Road exchange, and sensational stories of crytocurrency millionaires buying Tesla cars with their virtual coins. And yet the most enduring innovation may in the end just be the idea of the blockchain.

Blockchain was the innovation of the elusive inventor of Bitcoin, Satoshi. It was a solution to the computer science paradox known as the double-spend problem. Namely, how do you send and receive money online without the need for a trusted third party to ensure the same digital credit standing in for the amount being exchanged isn't being spent twice.

Satoshi's approach was to create a ledger of all transactions owned and monitored by everyone but ultimately controlled by no one. You access and update the ledger to confirm each digital credit is genuine.

If you think more broadly, the blockchain represents a trusted, shared transaction record that allows machines to own and exchange value without human intervention.

In the future, we might use the same approach to build other exchanges – from spare cloud computing cycles and shipping capacity to things that are harder at present to quantify, like trust and reputation.

What other industries are at risk of being disrupted by the replacement of intermediaries with a distributed transaction record?

Botnet

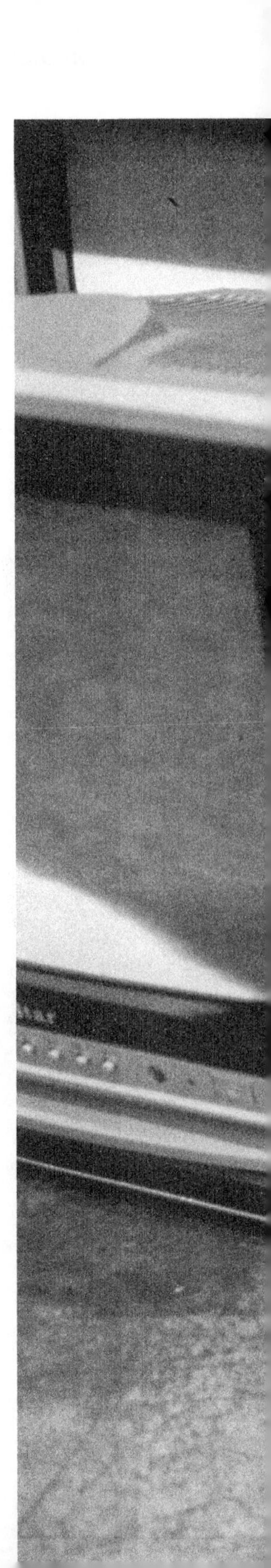

A botnet is a network of programs running on compromised computers and controlled as a group without the owner's knowledge.

WHAT IF THE future of crime were online?

GameOver Zeus first started circulating in September 2011. It began with a professional looking email that contained information about an unpaid invoice. Opening the email would trigger the program, and the next time you visited your online bank, the system would eavesdrop on your online session. The program was smart enough to bypass two-factor authentication and displayed a fraudulent banking security message for transaction authorization. Additional security questions were relayed to the user and, upon receipt, the hackers would steal the user's money.

GameOver Zeus was a botnet: a sophisticated collection of viral, web connected programs designed to allow the bot master to perform illegal actions. Typically, these botnets are instructed through a command and control server. What made GameOver Zeus particularly deadly was that it was designed around an encrypted peer-to-peer communication protocol that made it very difficult to detect and shut down. These peers acted like a giant network used to distribute updates, configuration files, and send stolen data.

At the time of the FBI's takedown of the botnet, it was estimated that GameOver Zeus had been responsible for over $100 million stolen from banks, businesses and consumers worldwide.

Botnets are the digital equivalent of a criminal network. They spread wildly and can remain undetected for long periods of time. They are also available for hire – whether your service of choice is sending spam email, mining Bitcoins, click fraud or participating in denial-of-service attacks.

In the case of GameOver Zeus, the botnet was also used to distribute CryptoLocker software, which when activated, encrypted a user's files with public-key cryptography, with the private key stored only on the malware's control servers. If you didn't pay using Bitcoin or a pre-paid cash voucher by the deadline, the malware threatened to delete the private key.

Which is a greater source of vulnerability – your systems or how you use them?

Carbon Nanotube Computer
Citizen Science
Clickbait
Cloud
Co-creation
Cohort Effect
Collaboration Map
Computational Design
Computational Thinking
Consumerization
Contextual Computing
Crowdfunding
Crowdsourcing
Cyptocurrency

Carbon Nanotube Computer

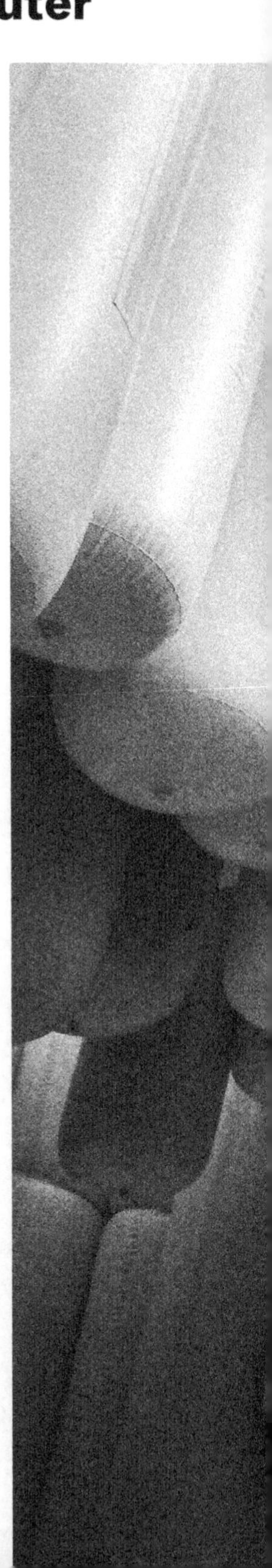

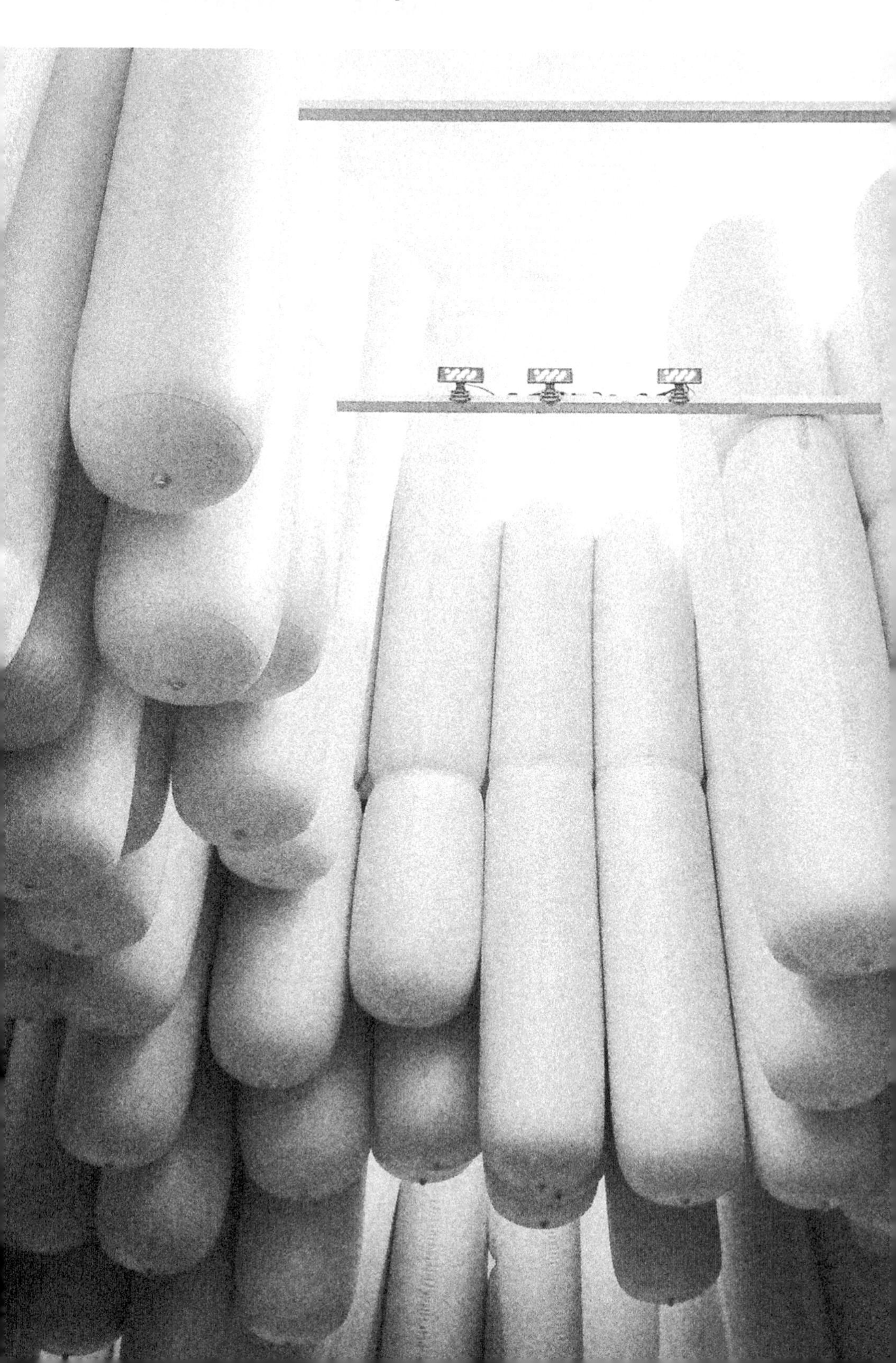

A carbon nanotube computer is a system built using carbon nanotubes rather than silicon transistors.

WHAT IF WE could build computers using nanotechnology?

Moore's Law, a prediction from 1965 that the number of transistors crammed into circuit would double every two years, has an expiry date. The problem is scale. The latest chips from Intel have silicon transistors with features as small as 14 nanometers. Theoretically you can have a feature as small as a single atom, but before you reach that point – at about 7 nanometers, things get weird. You leave the conventional world of classical physics and open a portal into the trippy reality of quantum physics. That's bad news because by 2020, in order to keep up with Moore's Law, the industry will need to be down to five nanometers.

An alternate idea is whether we can ditch silicon, and build computers using transistors made of carbon nanotubes. In 2013, a team of researchers at Stanford University built the world's first computer prototype based entirely on carbon nanotubes. They named it Cedric.

Prof. Subhasish Mitra, one of the project's co-leaders, compared the capabilities of their carbon nanotube processor to the original Intel 4004 released in 1971. Slow, with just a single bit of information, Cedric could only count to 32. But what made the project interesting was how the team overcame many of the challenges of growing carbon nanotubes straight enough to fit on a wafer.

Sidestepping Moore's Law is not the only advantage of carbon nanotubes. It is believed that these next generation computers may be far more energy efficient than traditional silicon-based systems. With a greater ability to dissipate heat, they may avoid the problem of today's computers that are effectively speed-limited by their design.

If a smartphone today has more computing power than all NASA had when it put a man on the moon in 1969, what new capabilities might a standard consumer device have by 2020?

Citizen Science

Citizen science is the collection of data, development of technology and participation in research by ordinary people.

WHAT IF AMATEURS were essential to scientific research?

Citizen science is not exactly a new idea. There is a long history of amateur or well-heeled hobby scientists' discoveries proving monumental – Sir Isaac Newton, Charles Darwin and Benjamin Franklin included. However during the 20th century, research became largely institutionalized, driven by universities, government and corporate research laboratories. Part of the reason was the sheer cost of doing detailed investigation in many fields, but another was the rise of the cult of academia and the formalization of published research.

The Internet has started to change things. It is easier now for ordinary people to volunteer their time to participate in the scientific process. Like an elaborate CAPTCHA scheme, NASA's Clickworkers was a program that allowed the general public to assist in the identification and classification of images like craters on Mars taken from the Viking Orbiter.

Another example is the use of gaming. Cell Slider is a web-based game that recruits amateur pathologists to differentiate breast cancer subtypes. Foldit, on the other hand, attempted to predict the structure of a protein by taking advantage of people's puzzle-solving abilities and having people play competitively to fold the best proteins. The results from Foldit have been applied to research in AIDS, Cancer and even Ebola.

Sometimes a more direct approach is required.

Joi Ito, who is the head of the MIT Media Lab, tells the story of how on the day of the devastating Japanese earthquake in 2011 and subsequent nuclear reactor meltdowns, he went online and organized a group of volunteers. They called themselves Safecast, designed their own Geiger counters, collected 16 million data points and created an app that displayed radiation in Japan and other parts of the world. Safecast is now the largest open dataset of radiation measurements.

Is there an aspect of your research and development that you might crowdsource to the general public?

Clickbait

Clickbait is a style of web content that utilizes sensationalist headlines to attract click-through and encourage viral forwarding over social networks.

WHAT IF SOCIAL media were destroying quality content?

Upworthy is a curious beast. Launched in March 2012 it pulls in over 60 million readers each month while dominating social media networks. Its headlines use all of tabloid media's greatest hits – the classic attractors of sex, celebrity and miraculous cures – but it also claims to be a "mission-driven media company" dedicated to raising awareness of social issues and good causes. Go figure.

The trend is spreading across the media landscape. Rival Buzzfeed adopts a similar approach, as do the online media publishing divisions of stalwart publishers like Time and Forbes magazines. It is not by chance that the headline "9 Out of 10 Americans Are Completely Wrong About This Mind-Blowing Fact" gets nearly 7 million page views, it is largely because of human psychology.

Eric Jaffe, writing in the magazine Fast Company, argues that Upworthy crafts headlines designed to create a "curiosity gap". It's an idea that builds on the work of George Loewenstein at Carnegie Mellon University. In the mid-1990s, Loewenstein observed that "the curious individual is motivated to obtain the missing information to reduce or eliminate the feeling of deprivation." Clickbait headlines typically set up the premise of a story or an outrageous idea, and require a reader to click through in order to complete the pattern in their brain.

Analytics are a large part of Upworthy's approach. The editorial team scours the web for topic ideas, writes multiple headlines for a sample audience, A/B tests the results, chooses the best performing versions and then heavily distributes the resulting story on social media. It may be sensationalism, but it is of the statistical breed.

Some see Clickbait as an inevitable consequence of how audiences consume media. Social networks have increased the distribution of content, but have also encouraged much shorter attention spans. In a fragmented world, media has to scream for attention. More challenging, perhaps, is the role of data in traditional journalism. Should page views be the ultimate determinant of what kinds of stories are written, and which journalists get hired?

When live audience data meets the newsroom, is tabloid journalism an inescapable conclusion?

Cloud

The Cloud is the network of servers that together provides computing and software as a service rather than as a product.

WHAT IF THE Internet were the operating system?

It's fair to say that I was an early adopter. Even before the Web, in the early 90s I was already an enthusiast, accessing bulletin boards with an acoustic coupler modem so slow that it loaded pictures one line of pixels at a time. From that perspective, it was inconceivable that one day we would be using the same network to stream high definition movies. Even stranger was the idea that it would become the way we would also access software.

At a certain point in time, we stopped thinking about the Internet as pipes, routers, and computer servers – and instead as a collection of services. This shift in thinking was captured by the metaphor of the Cloud, a platform that now runs much of the world's applications, can perform complex calculations on demand and provide storage for both our personal and enterprise content.

For large companies, the Cloud represented a big change in thinking. Rather than running your own on-premise hardware and software environments, you had to contemplate hosting your data with other people – your information co-mingling on the same platform, although securely partitioned. For that reason, some businesses – especially those in heavily regulated financial or public services – explored a hybrid model that incorporated a private cloud to reduce their potential vulnerability to outside interference.

The idea of the Cloud is more than just an evolution in enterprise architecture. It also represents a challenge to the role of IT in business. As the Cloud brings higher levels of automation, IT teams no longer have to devote all of their resources to installing patches and maintaining hardware, but can instead turn their attention to figuring out how to help the business develop a competitive advantage through technology.

Strategic benefit is the real point. Generic technology is no more a source of advantage than running water or electricity – it is a utility. The opportunity for companies lies in using the flexibility and scale of a cloud-based world to design new experiences for customers, as well as pushing the boundaries of how employees work, communicate and collaborate.

How will the Cloud help you challenge your own business models or disrupt someone else's?

Co-creation

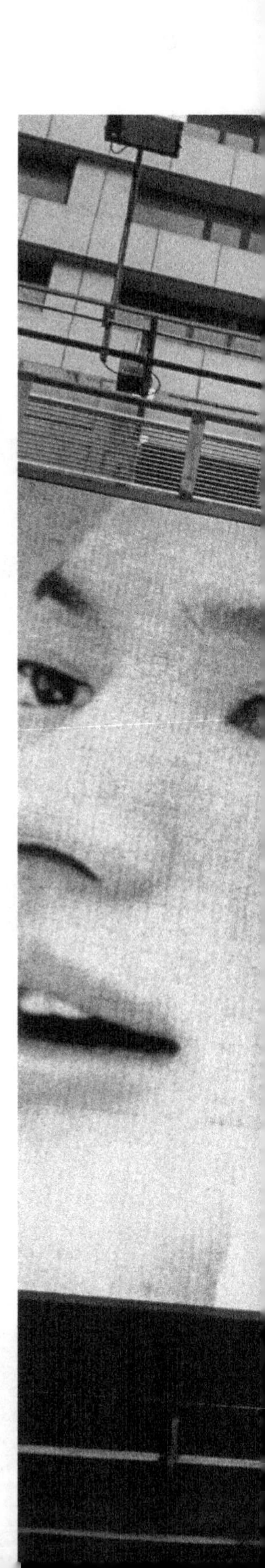

Co-creation is the process by which companies engage their customers in the design and delivery of their products and services.

WHAT IF YOUR customers were responsible for design?

Co-creation is the idea that design is no longer something that happens without the participation of the customer. Originally conceived by C.K. Prahalad and Venkat Ramaswamy in their Harvard Business Review article, "Co-Opting Customer Competence", co-creation has again become a big idea due to the rise of digital platforms.

Consider a company like Amazon. Their platform is their product, and yet no customer who visits their home page sees the same screen. As you buy products, share information, and interact with the service – the platform re-configures automatically around your interests. The same applies to Netflix, the online entertainment service where recommendations mirror behavior.

Co-created companies don't have to be digital. Companies like Starbucks and Best Buy have created engagement platforms that solicit consumers for ideas on how to change their business. And yet not all digital companies always follow the co-creation model. Apple for instance is in many ways a closed model of innovation. Their designers work in absolute secrecy, furnishing the world annually with a new smartphone model that incorporates a highly individualistic model of design. That said, Apple embraced co-creation to launch its App Store ecosystem – generating over $1 billion in revenue for its partners within two years.

The future of co-creation lies in platform enablement. At 18, Ben Kaufman persuaded his parents to take a second mortgage out on their home so that he could start an iPod accessory company called Mophie. Though successful, the experience convinced him that coming up with ideas for products was not as hard as getting things made and distributed. Quirky was the result, a platform that allows people to come up with ideas for product, and if they meet enough pre-orders from the community, they can share in the resulting revenue when the products go on sale at major retailers.

If your customers had the power to change one thing about your business, what would it be?

Cohort Effect

The cohort effect refers to the impact made by generational groups who have grown up with a common set of experiences.

WHAT IF YOUR kids were in charge of your business?

I'm both fascinated by and suspicious of claimed cohort effects. Marketers love trends, and the idea of generational groups is like catnip to those that want to justify their own crazy ideas. Certainly, there are no shortage of generations – Generation X, Generation Y, and now the Millennials. The rules for entry into these categories are as wide as they are somewhat arbitrary, while descriptions of their expected behaviors often read like horoscopes.

That said, cohort effects are still very real. You just need to pick the right parameters. In my view, cohorts make the most sense when you define them based on exposure to disruptive technologies. Put a tablet in the hands of a baby, and you can imagine how those brand new neurons might react to the incredible stimulation. Entirely new pathways and connections will be built through those experiences that will shape their development in later life.

One of the first cohorts I tracked and wrote about in my book Futuretainment was the group born after the birth of the Web in 1994. I called them the Naturals, because in a sense they were the generation who grew up not having to learn the digital world. From the way they entertained themselves to how they communicated, looked for a job or managed their finances, the Web was their default choice.

The gap between technology cohort-forming events is now getting shorter. You had the smartphone generation born after the introduction of the first iPhone in 2007. Soon after, you had the first group of children born with iPads used as their virtual babysitters. Later, you will see cohorts emerging when wearable technologies become mass market, and then those who have personal robots as family members. Further out still, but potentially the most disruptive, will be the first cohort of children to be born with artificially modified genes. Their neural re-wiring may actually be a design feature rather than an effect, conceived on a computer screen well before they are even born.

If your kids joined your organization, what would be the first thing about your use of technology that they would find puzzling?

Collaboration Map

A collaboration map is a model of how a company's employees work together to achieve their goals.

WHAT IF THERE were a smarter way of understanding how people did their work?

People working together sounds like a good idea. That's why CEOs talk about collaboration in their vision speeches, technology vendors try to sell it to you as a package, and every ambitious manager claims to do it. And yet, for many organizations, it remains an elusive ideal.

Eric Schmidt, chairman of Google put it bluntly. He said, "When you say collaboration, the average 45-year-old thinks they know what you're talking about: teams sitting down, having a nice conversation with nice objectives and a nice attitude."

The problem is that collaboration is more than the sum of its moving parts. You can set up instant messaging, email, video conferencing, an internal social network and even a few telepresence robots for good measure – but designing the right collaboration culture is an anthropological rather than technological problem.

More important than the tools you use is mapping the communication modes that exist inside your business. Ask yourself: how do your best people work together? How do they resolve the challenges of different cultures, remote workers and departmental politics?

McKinsey & Company described in an article titled "Mapping The Value of Employee Collaboration" how one of their clients, a leading biotechnology company, relied on sharing best practices among quality control engineers. Using network analysis, they knew which engineers took part in interactions that generated time and cost savings. When mapped to economic value, the company could identify just how much money the collaboration breakdowns were costing them, and where they needed to invest further in tools and training.

Collaboration is a cultural construct, but also a coordinative one. As people's roles become more fluid and teams more globally distributed, having a map to navigate the future of work will only become more important.

Are the best teams at work those that are formally created, or those that emerge spontaneously in response to a problem or opportunity?

Computational Design

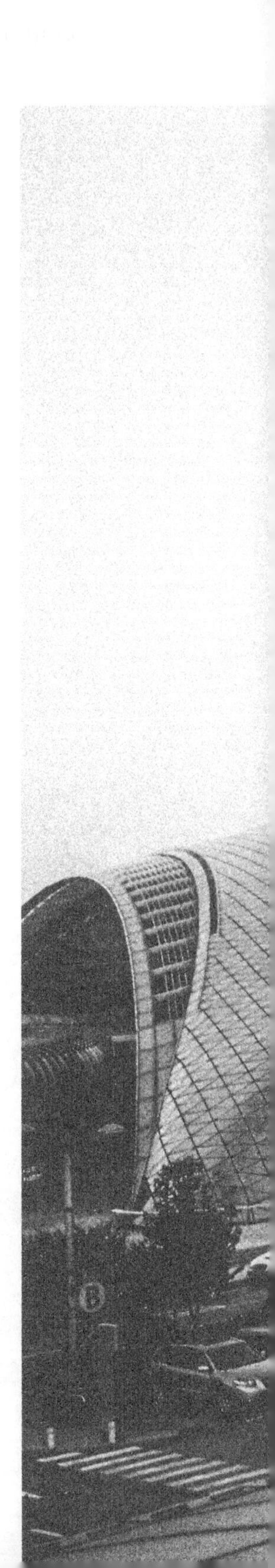

Computational design uses algorithms to generate optimal solutions based on parameters set by a designer.

WHAT IF THE world's best designer were an algorithm?

When some of the world's biggest technology brands – Google and Amazon in the US, and Alipay and Tencent in China – started planning new office spaces, they turned to a Seattle-based firm. That firm, NBBJ, is known for taking an unusual approach to design.

NBBJ describes its methodology as computational design. They use algorithms and computer models to simulate how a building's occupants will actually use a space. Their software links geometry with data to address specific problems such as the kinds of views available from different offices, and the best way to foster collaboration through visible sight lines.

According to Paul Audsley, who is NBBJ's Director of Design Technology, design computation allows the development of "intelligent, flexible building models that provide instant visual feedback along with key supporting data to help us 'prove' our design concepts at the earliest possible stages."

The use of algorithms in building design is part of a broader shift in manufacturing that will impact a wide range of everyday products and industrial objects. Autodesk for example, have set up a research division called Project Dreamcatcher to create a goal-directed design system. The system will allow designers to set certain parameters – such as material type and performance criteria – before evolutionary algorithms on a Cloud platform are given the opportunity to create thousands of valid design options and recommend the best-performing versions for further consideration.

Computational approaches to design will have a potentially disruptive impact on manufacturing. New tools will allow designers to generate optimal solutions to complex problems, leading to the creation of entirely new types of objects. Humans will not be replaced in the design process – but whether it is a car, a chair or a HVAC system – software will run through thousands of simulations before suggesting potential solutions based on the rules and constraints set by the designer.

If you used an algorithm to design your next office building, what characteristics would you optimize for?

Computational Thinking

Computational thinking is a way of solving problems based on principles from computer science.

WHAT IF YOU could teach children to think like computers?

You might say that our entire modern era of computing began with the thought experiment of a bright young man who developed a rather machine-like way of seeing the world.

In 1936 when he was only 23, Alan Turing wrote a paper in which he proposed what became known as a universal Turing machine. His thought experiment involved an infinite tape divided into squares like a child's exercise book. The tape was the input, working storage, and output of the system. A tape head could read the current square, write a symbol, and move left or right one position. The tape head could also keep track of its internal state, and followed instructions as to what to do next. The genius of Turing's model, despite being proposed long before computers actually existed, is that it can still represent the logic of any computer algorithm in use today.

Learning to think like a computer may be more important than knowing how to program one. Jeanette Wing, Head of Computer Science at Carnegie Mellon University, wrote an influential paper in which she argued that computational thinking was the new literacy of the 21st century. Her view was that core skills like being able to see patterns, generalize from specific instances and formulate problems in a way that would enable computer tools to aid in solutions – was not just relevant for those working in technical sciences, but for all students.

With a knowledge of computational thinking, a medical researcher might find more scalable ways to conduct patient trials or ensure drug designs are less likely to result in drug-resistant strains of diseases. Architects and city planners could create models for how people might use a building, and better project the impact of growing urban density. Even artists and writers could leverage data and algorithms to create radical new works that provide critical insight into the contemporary condition.

Whatever profession the next generation decides upon, data and its analysis, as well as an appreciation of the issues of scale and complexity, are likely to remain very relevant themes.

If you could influence the educational curriculum to create graduates better suited for your company, what would you change?

Consumerization

FENDI

Consumerization refers to the influence of consumers on the design of technologies in the enterprise.

WHAT IF WE were to design technology for work based on how we use technology at home?

It started with the iPhone. Despite decades of serious development and millions in funding, most people's experiences of technology at work had not been particularly inspiring. Devices were clunky, interfaces non-intuitive and IT policies restrictive. If IT called you, it was because your inbox was too full or you had been looking at the wrong kind of websites. But in your personal life, it was a very different story. Gorgeous devices, delightful web-based tools, and the freedom to develop a much more fluid digital work style.

Not surprisingly, as the next generation started to join the workforce, they brought not only their devices but their expectations with them. Employees with good technology at home expected to be able to use it at work too. Traditional IT was forced to embrace a policy of "bring your own device" and allow employees to connect their own phones and tablets to their work networks.

However, consumer orientated companies like Google and Facebook had also invested heavily in the user experience, and that influenced how people wanted applications to operate in the enterprise as well. Employees outside the technology department began to use their own cloud-based applications without official approval – leading to the rise of "shadow IT".

Consumerization has its own set of challenges. Some companies take the opportunity to require their employees to be accessible by phone without paying for the costs of doing so. A recent California Court of Appeals ruling held that employers must now reimburse a reasonable part of employees' cellphone bills when use of their personal phone is required to do their work.

The greatest challenge of Consumerization was and remains security. With the complexity of personal devices and the possibility of sensitive information being stored on a variety of cloud-based services, IT teams will need to shift their focus from securing devices to securing users.

Does allowing people to bring their own technology to the office create better productivity, or a worsening balance between work and life?

Contextual Computing

41号

Contextual computing utilizes data about your surroundings in order to present you with relevant and actionable information.

WHAT IF YOUR devices knew what you wanted, even before you did?

In the near future when you walk to your car in the morning, your GPS will show you your destination before you even touch the screen. It won't be pre-programmed or a feat of mind reading, but a simple deduction based on your last location, the day of the week, the time, your calendar, your email inbox, or what you last posted on social media. We think of these as distinct fragments of data, but when taken together – they are your context.

The idea of contextual computing was first described by Georgia Tech researchers Anind Dey and Gregory Abowd. They imagined context-aware devices as ones which, when given information about their circumstances based on rules or an intelligent stimulus, could react accordingly. In their view, this would ultimately allow a greater richness of interaction. When humans talk with humans, they are able to use implicit situational information or context to increase the conversational bandwidth. Computers need to do the same.

Google Now, for example, accesses both your personal data and information from your smartphone's sensors, to provide you with more nuanced information – even when you are not expecting it. Larry Page, one of Google's co-founders, observed in a fireside chat with venture capitalist Vinod Khosla that even though most of Google's products were structured around the queries people made, there were times when "maybe you don't want to ask a question. Maybe you want to just have it answered for you before you ask it."

Contextual computing will become even more important when wearable technologies, often without large screens or input devices, become mainstream. Retail stores, offices, airports, even different rooms in your home, are all full of potential contextual data, that could trigger applications and push relevant information to you when appropriate.

"The most profound technologies are those that disappear," said Mark Weiser in 1991, chief scientist at Xerox PARC and an early visionary of the idea of ubiquitous computing. "They are those that weave themselves into the fabric of everyday life."

If you were to design a digital application that worked without any kind of user input, what would it do?

Crowdfunding

Crowdfunding allows entrepreneurs to raise funds directly from the public, without resorting to venture capital, banks or other traditional sources of finance.

WHAT IF COMPANIES raised money from customers rather than financiers?

The idea of funding projects with contributions from potential customers is not entirely new. In 1713, the poet Alexander Pope decided to translate 15,693 lines of Homer's ancient Greek epic, the Iliad into English. With six volumes in total and each volume taking a year, he knew he would need some help. In return for their name in the acknowledgements of the book, 750 people pledged two gold guineas each to support his project.

Crowdfunding platforms have multiplied in recent years. According to estimates by Fundable, by the end of 2014 crowdfunding was estimated to add at least 270,000 jobs and inject more than $65 billion into the global economy.

Early crowdfunding campaigns on platforms like Kickstarter were based on rewards and benefits. Projects ranged from smart watches to high tech drink coolers, from ad campaigns to mobile phones. By March 3 2014, the company had reached $1 billion in pledges. More than half of that amount was pledged in the previous 12 months alone.

For inventors or startup companies, crowdfunding platforms provide a valuable way to get validation from customers as well as build a community of beta testers. For example, backers of Star Citizen, the PC space game from Chris Roberts, received early versions of the software and the ability to shape its development. The game broke records with $51 million in pledges.

Of course not everyone is happy when a company like Oculus Rift raises $2.4 million in donations only to sell for $2 billion to Facebook 18 months later – leaving the crowd with nothing but some nice t-shirts. That's why the new trends in crowdfunding are equity or credit-based models. One such company, Fundable, lets startups raise equity funding from investors. A similar player, CircleUp, targets consumer goods businesses with annual revenues of at least $1 million, partnering with major brands like Johnson & Johnson, P&G and General Mills to provide data on product development trends.

If your next product were to be crowdfunded, how would that change your approach to design and positioning?

Crowdsourcing

airtel
OSHO GLIMPSE
BOOK HOUSE
WELCOME FOR OSHO BOOKS
CDs, DVDs AND MEDITATION
WELCOME TO BOOK HOUSE
BANARAS SAREE FACTORY
airtel
airtel
airtel
airtel
airtel
airtel
श्री बालाजी सिल्क साड़ी
FACTORY

Crowdsourcing is the outsourcing of internal projects to external groups of people.

WHAT IF A group of strangers knew more about the solution to your company's problems than you did?

Sometimes you have to assume that the best people don't work for you. From logo design to complex data challenges, crowdsourcing has become a growing tool for companies looking for external inspiration.

Crowdsourcing works because it generally involves the breaking down of large projects or problems into smaller tasks that can be completed by groups of independent professionals, experts or talented amateurs.

There are two ideas at the heart of crowdsourcing. The first is that the right expertise for the problem at hand may often lie outside your organization. This can be for both cultural as well as economic reasons – smart people often like more personal autonomy and a high-paying freelance lifestyle rather than having a fixed job. But it may also be that you have been looking in the wrong place altogether for your solution.

Anthony Goldbloom, an Australian economist, founded Kaggle based on the assumption that companies and governments have lots of data, and lots of problems that data could solve. What they don't have are the right kinds of experts working for them. In one competition, held by the world's leading space agencies, it took one week for Kaggle to solve the problem of measuring the way dark matter distorts the images of galaxies. Interestingly, the solution came from a glaciologist.

The other idea that makes crowdsourcing interesting is that it taps into the complex ability of crowds to generate innovative and effective solutions to problems. Wikistrat, for example, is a platform that focuses on predicting geo-political events and potential flashpoints for conflict. They use a "closed crowd" of subject experts and a model of collaborative competition that rewards participants for the breadth and quality of their contributions. The diversity of opinions in crowdsourced collaboration is a good way of avoiding groupthink.

How do you integrate great ideas from outside your firm, without having them defeated by the political agendas of those within?

Cryptocurrency

DOUBLET

A cryptocurrency is a form of money that relies on digital cryptographic techniques for its security.

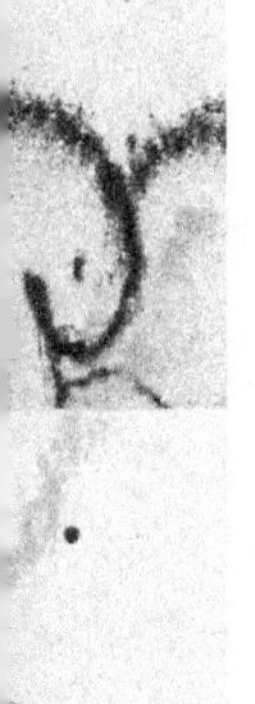

WHAT IF THE future of money were entirely digital?

The arrival of a beautifully printed and engraved letter from Japan on quality card stock from the court-appointed bankruptcy trustee was a paradoxical reminder of a painful truth. I had been a victim of a cryptocurrency collapse. Some months earlier, I had invested a modest amount in Bitcoin through the largest exchange of the time, Mt Gox. Then in February 2014, the exchange announced it had lost nearly $473 million of its customers' funds, apparently due to a massive online heist. This was equivalent to approximately 750,000 Bitcoins or about 7% of all the Bitcoins in existence.

The problems at Mt Gox were not precisely the fault of Bitcoin but rather the complexities of dealing in a virtual currency. If I had been more paranoid, I would have elected for cold storage – the practice of taking your coins offline by storing their codes on a piece of paper locked in a vault. Safe perhaps from the reach of hackers who might steal them, but not exactly a convenient place to spend them either.

Bitcoin is just one of many cryptocurrencies vying for survival. Nor will it be the last. Although different specifications govern their creation and distribution, cryptocurrencies generally share some similar characteristics. Unlike fiat money, which is issued and printed by governments in centralized banking systems, cryptocurrencies lack official backing. Instead, they rely on peer-to-peer technology, which enables all functions including currency issuance, transaction processing and verification to be carried out collectively by the network. Miners, whose computers are required to solve progressively more difficult algorithms, are rewarded for their effort with coins. The total number of Bitcoins that can ever exist is 21 million, a level likely to be reached in 2140.

Governments may not back Bitcoins, but regulators are already moving in. In the US, the IRS has ruled that Bitcoins should be treated as property, and thus be subject to capital gains tax. Longer term, central banks have acknowledged that the widespread adoption of cryptocurrencies may challenge their ability to influence the price of credit and other traditional tools of monetary policy.

What would you have to believe in order to be comfortable doing business in a virtual currency?

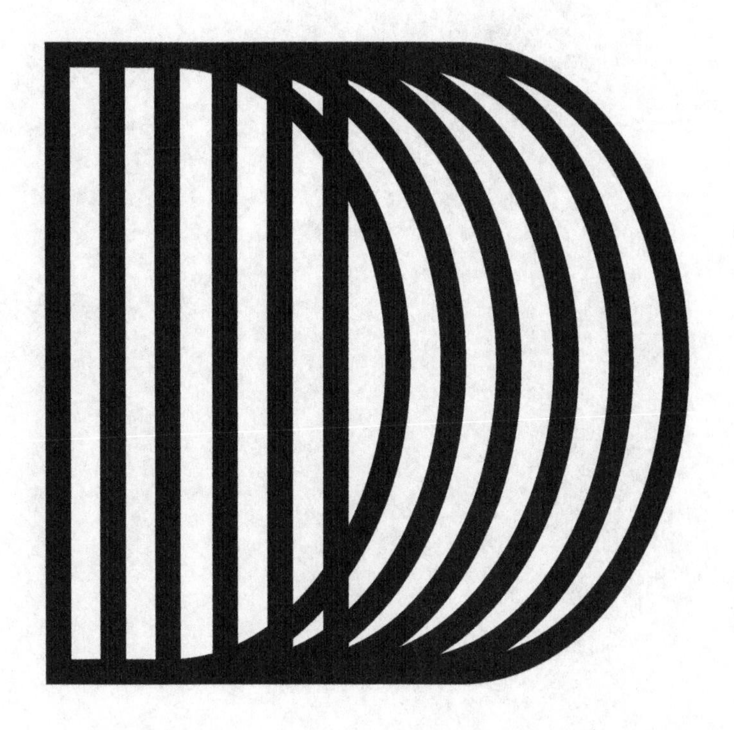

Data Profiling
Deep Web
DevOps
Digital Uprising
Divergence
Drone

Data Profiling

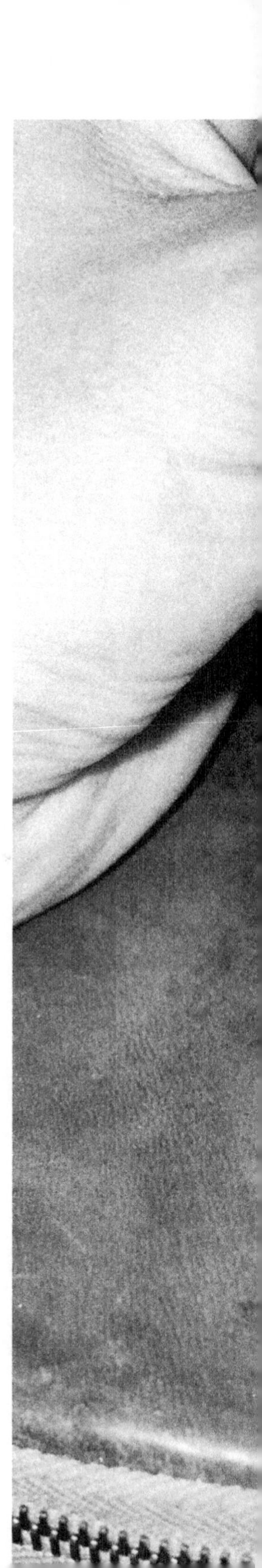

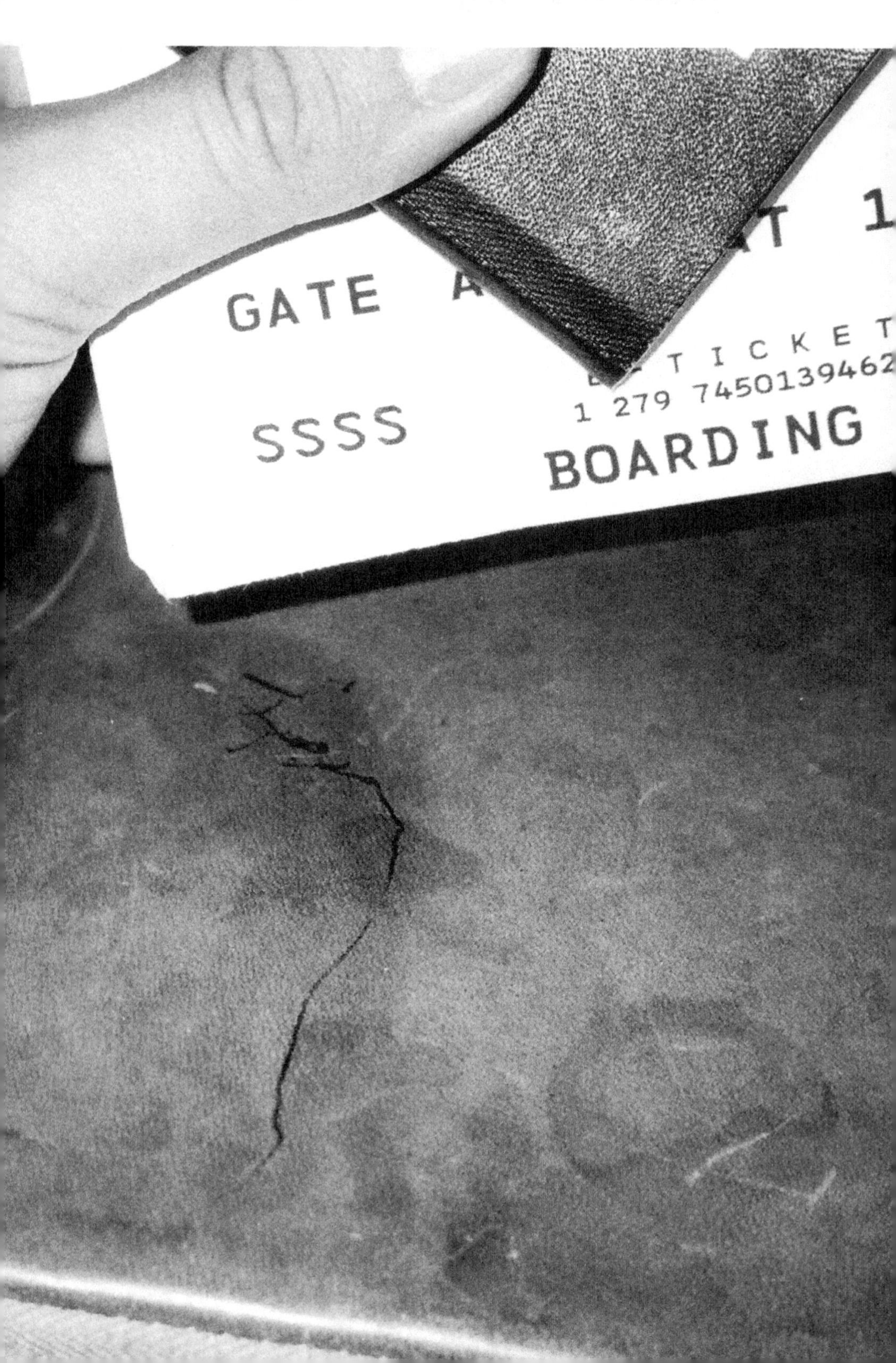
GATE
SSSS
1 279 7450139462
BOARDING

Data profiling is the automated processing of an individual's personal data in order to predict their personality or actions.

WHAT IF THE greatest risk to your privacy were what people think you might do?

Every time I board a flight heading to the United States, a special code appears on my boarding pass, SSSS. At first I thought this was some kind of frequent flyer reward, but suffice to say, I was wrong. If you get that code, it means you are on a selectee list, and your only reward is being detained for invasive searching each and every time you fly. Somehow my travel patterns, behavior or just my name had triggered an algorithm that identified me as a high-risk traveller.

Data profiling is not a new trick. When you fill out a warranty form, answer a call from a telemarketer, join a gym, or sign up for a credit card, you update a series of consumer databases that are cross-referenced and traded. In one controversial case, Metromail (now Experian) actually used prisoners to enter personal information into computers, leading to a stalking case where a convicted rapist harassed a woman based on the information she had entered in a survey.

That computers are replacing prisoners in the collection and analysis of personal data might reassure some people, but it raises a different kind of problem. As government, border police, insurance companies and other financial institutions start to use algorithmic profiles to make decisions, individuals will receive increasingly less context for why they are being treated as they are. This classic Kafkaesque dilemma is made worse by the inevitability that data mining and pattern recognition programs will reach a level of complexity such that even the people administering them will no longer completely understand how they work.

There is a fine line between invasive data profiling and personalization. On one hand, consumers and citizens are comfortable with a certain level of data analysis that leads to better customer experiences and safer cities. The real problems begin when the systems we engage with are no longer transparent, let alone have privacy settings.

In the near future, will privacy be a luxury for just the rich and paranoid?

Deep Web

USE STAIRS
UNLESS OTHERWISE
DIRECTED
DO NOT
DISTURB
PLEASE

The Deep Web is the hidden part of the Web not indexed by major search engines.

WHAT IF THE Web were only a tiny piece of the world's actual digital resources?

When police stormed a public library in San Francisco in October 2013 and arrested the pale, unassuming man quietly typing on his laptop, they also drew back the veil on a hidden world. The man was Ross Ulbricht, better known by his associates as the Dread Pirate Roberts. Ulbricht had founded "The Silk Road", an eBay for everything from drugs to assassins, stolen goods to hackers for hire. The turnover of this underground network was estimated to be $1.2 billion with over a million customers.

The Silk Road existed within what is known as the Deep Web. Most people spend their time in the Surface Web, made up of content indexed by standard search engines. Yet, the Surface Web is understood to comprise only 4% of the actual information out there. The rest is stored in academic research libraries, proprietary databases and corporate intranets – all beyond the reach of conventional search engines. In one way, the Deep Web begins at the limits of our ability to interpret and classify information.

Accessing the Deep Web requires a special browser, such as TOR (originally an acronym for The Onion Router Project). TOR bounces your signal through an endless network of host PCs until it's almost untraceable, obscuring your location and what you are looking at. Ironically, TOR is not some rogue technology but was developed 11 years ago with funding from the US military.

With growing fears of state based surveillance, a wider demographic of web users have started experimenting with tools and sites that protect their privacy. For example The New Yorker Strongbox is a secure communications channel you can use to submit messages to their editorial staff.

The Deep Web reflects the anti-establishment, freewheeling spirit of the early days of the Internet. At the time of his arrest, Ulbricht's LinkedIn profile said he was working on "creating an economic simulation to give people a first-hand experience of what it would be like to live in a world without the systemic use of force" of the kind imposed by "institutions and governments." A dangerous idea indeed.

How much of our knowledge is subject to the whims and algorithms of those who organize it?

DevOps

DevOps is the melding of software development and IT operations to facilitate rapid technology innovation.

WHAT IF YOU could approach software development like a Web giant?

For most companies, technology innovation is hard to get right. On one hand, you have developers making software. They like change, innovation and doing things differently. That's a very different mindset from your other group, IT operations. Their job is to keep things running, avoid outages and maintain stability. Suffice to say, they don't particularly enjoy change.

Rather than a new set of technology tools, the DevOps movement is a cultural philosophy designed for a world where companies have to be agile enough to constantly deliver new and exciting experiences for their customers. Your next release has to be ready for testing, just as the last release goes out. The reason for that is simple: fast technology cycles drive future growth.

Although greater communication and shared goals help bridge the divide between developers and operations people, companies can still experience growing pains. More frequent software updates mean shorter test cycles, so sometimes you have to use your real customers as live test subjects. Avoiding embarrassing and public failures means ensuring that you are actively pushing information from customer care and sales agents to your product teams, to catch problems before they become disasters.

One of the other factors driving change in IT is the fact that the underlying infrastructure is becoming too complex for human beings to manage. Greater levels of automation are replacing many traditional operations, while the next generation of application developers – working in cloud environments – are also starting to build testing and quality assurance into their workflows.

The ultimate goal for any technology-driven company is continuous delivery – the idea that as soon as a feature has been completed, it can be automatically rolled into production. Amazon claimed in 2011 to make changes to production every 11.6 seconds on average, while Facebook releases twice a day.

Are your technology teams up to the challenge of keeping up with your customers?

Digital Uprising

から追放せよ
権を
行動委員会
敵朝小沢
政界

A digital uprising is a civic revolution facilitated by social networking platforms and mobile devices.

WHAT IF SOCIAL media were capable of challenging the balance of power?

When protestors in the Ukraine woke on a Tuesday morning in January 2014, they were greeted with an ominous message: “Dear subscriber, you are registered as a participant in a mass disturbance. ”The message was most likely delivered by a “Stingray” device, a fake cellphone tower pioneered by US police.

The benefit of social media in digital uprisings is no longer as clear as it was during the early days of the Arab Spring, when personal technology provided a radical and resilient way to rapidly spread news, coordinate protest and circumvent unprepared authorities. The events that unfolded in Tunisia, Egypt and Turkey were presaged by earlier uprisings, first in 2008 in Seoul and then later in 2009 during the Iran elections.

“It is a world just for us,” explained the Saudi teenager to me as he handed me his phone. I was speaking at a conference in Riyadh, and as I looked through his friend’s profile pictures on Twitter and Facebook, I realized that Saudi youth actually lived in two worlds. The public one, with its formal codes of conduct, and a digital one, where anything was possible.

Saudi Arabia, with over 4.8 million active users, has the world’s highest penetration of Twitter users. Twitter has become a kind of pressure valve for the next generation. The phrase, “the salary is not enough”, referring to Saudi public employee pay rates, become the 16th most popular Twitter trend in any language with 17.5 million tweets in two weeks. The most followed account in the Kingdom is @mujtahidd, a mysterious satirist of the ruling family. And yet, despite all of these provocations, Prince Alwaleed bin Talal, chairman of Kingdom Holding Company, is one of the largest investors in Twitter.

What can be used to challenge authority can also be used as a tool of terror. Along with stolen Humvees and other US weapons, one of the most powerful Western tools appropriated by the Islamic State movement is social media, evidenced by their use of YouTube to globally recruit young Jihadis to join their cause.

As authoritarian governments become more digitally adept, what technologies will be most relevant for the preservation of civil protest?

Divergence

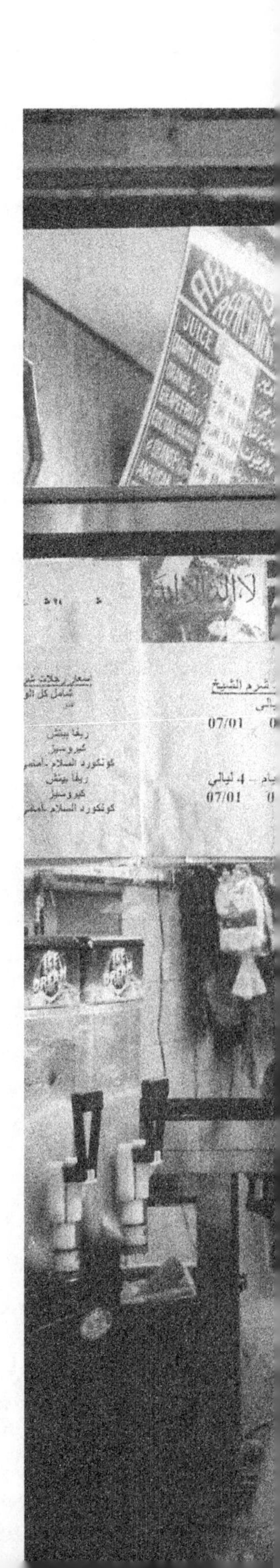

INTER NET CAFE
ON 2 ND FLOOR
PUSH
INTER NET
CAFE
ON 2 ND FLOOR
Telecard

A divergence is an evolutionary fork in technology, usually as a result of local culture or conditions.

WHAT IF THE future of the Web lay in the emerging world?

It's easy to look at today's US tech giants – Apple, Google, Amazon, Facebook and Microsoft – and imagine that the future of the Web is in their hands. But consider this. Of the world's 7 billion people, only 2.7 billion have access to the Internet today. The vast majority of the rest live in emerging markets like Africa, India and other parts of Asia.

Think about that. A billion new consumers joining the Web, bringing with them entirely new cultural perspectives. Every culture has its own unique twist on the digital world, whether it be the Turkish Web users who have embraced fortune-telling apps, the Indian consumers who love wedding arrangement sites, or the Hong Kong mourners who burn paper iPhones.

Cultural divergence will not only shape the platforms we are familiar with, but will also be the catalyst for entirely new business models. As my Nordic friends like to remind me, you really have to live in a country with as little sunlight every year as Finland in order to come up with an idea as perverse as a game involving suicidal birds.

Half a millennium ago, China led the world with its four great inventions of gunpowder, the compass, the printing press and papermaking. Then something happened. Historians call it "The Great Divergence". Essentially, the economic positions reversed during the 19th century and Western Europe and parts of North America become rich, while China became poor. The same thing could happen again, but this time in the digital ecosystem.

So despite the claims of altruism, today's technology companies are well aware of the potential and peril of tomorrow's new markets. Behind Facebook's internet.org mission to provide free Web access to the emerging world, and the Google Loon plan to use drones to deliver communications to remote locations – is a race to be first in the markets that may yet hold the key to the future shape of the Web.

If China's Baidu was the world's number one search engine rather than Google, how might things be different?

Drone

A drone is an unmanned vehicle controlled by onboard computers.

WHAT IF WE gave computers the ability to roam in the physical world?

Drones are more than mere robots; they represent a manifestation of computing systems in the real world. Whether you are a battlefield commander or a member of a crisis team searching for a lost plane, unmanned aerial, ground or undersea systems are essential tools that can enhance your situational awareness and provide previously inaccessible information.

However, there is no escaping the ethical dilemmas posed by military use of drone technology. For example, how much bounded autonomy do you grant a drone? If having a radio connection with an operator means being potentially detectable, should you give a machine the ability to make a kill decision based on pattern recognition?

War fortunately is only a small part of the future potential of drone technology. There are growing civilian applications from deep sea search, to emergency response and commercial delivery. Amazon's Prime Air concept may have been a publicity stunt, but it was also a powerful demonstration of the future of distribution. In remote or inaccessible areas, drones are likely to be a key part of emergency supply networks for medicine or essential goods.

One of the most promising areas for civilian unmanned systems is precision agriculture. Using drones for crop surveillance can drastically increase farm crop yields while minimizing the cost of walking the fields or airplane fly-over filming. With high-resolution sensors, the imaging from drones will also help decrease agricultural water and chemical use.

For now, the biggest challenge facing the drone industry is one of public perception. The COO of a major telecommunications firm told me recently that one of his clients, a mining business, had recently deployed autonomous mining trucks. It all was going well, with less accidents, higher efficiency, and better data. Everyone was happy with the robot trucks, except the truck drivers. The union was in revolt. What did they do? I asked him. He looked at me and smiled – they put crash test dummies in the drivers seat, he said.

For what kind of drone activities should a human always be in the loop?

Emoji
Enterprise Social Network
Enterprise Virtualization
Exascale
Experience Design

Emoji

Emoji is an iconographic language used as a substitute for words.

WHAT IF THE future of language were not verbal?

Here is how the story goes. In the 90s, the world's greatest phones were in Japan. While the rest of us made do with basic calls and text, Japanese flip phone users had built in cameras, music, newspapers and sophisticated location services. But as Japan's 80 million users started using picture messages to communicate, it placed immense strain on the network. To help reduce the load, the first emojis were developed in 1998 by Shigetaka Kurita, who was part of a team working on the Japanese mobile Internet service i-mode. The trend took off, and soon emoji became a national craze.

When Steve Jobs took his newly designed iPhone to Japan to find a local partner operator, he found a surprisingly skeptical audience. Japanese users were used to their own mobile culture. Softbank ended up agreeing to a deal – on two conditions. They would take a larger cut of the profits, and more curiously, the next iPhone had to include support for emoji. Not long after, emoji was integrated into Unicode, the industry standard that regulates the presentation of text across different software platforms.

Emoji is more than just LOLcat cuteness. Mobile communication is challenging. In linguistic terms, there are always two ways to read a message. The denotation and the connotation. Depending on context, there can be multiple connotations to one message, and without visual and verbal cues, it can be hard to know what someone actually means. An emoji helps provide emotional context.

Fearful parents observing their keitai-obsessed children's use of strange symbols and abbreviations might well wonder whether we are witness to the emergent rise of a pictographic universal language like the American Indian Hand Talk or the fabled original language spoken by humans before the fall of Babel. Unlikely perhaps for English and other languages based on phonograms, or symbols representing the spoken sounds we "hear" in our heads rather than concrete images. But it does raise the interesting question: how do you design visual operating systems that work across cultures and overcome the contextual challenges of mobile communication?

How would you market to a generation who no longer used words to express their feelings and desires?

Enterprise Social Network

An enterprise social network is the social graph of an organization as enabled by a social communications platform.

WHAT IF THE most important social graph was the one at your work?

Despite considerable investments in technology, most companies face two common problems. How do you find the right talent inside your organization with the skills you need, and how do you locate the documents and resources relevant to your project? Both may sound like simple challenges, but the lack of quality metadata means that simply searching for people or files will not help you.

Enterprise social networks are one strategy to address the problem. They are a kind of internal Facebook, providing a combination of internal blogs, wikis, chat and sharing tools to illuminate the relationship between people and content. They are based on the premise that the social graph inside organizations is as important, if not more valuable, than those we use in our personal lives.

Microsoft, for example, is now using signals from email, social conversations, documents, sites, instant messages and meetings to create what they called an Office Graph. The idea is to use this data to suggest related content and potential collaborators to users, based on the work they are doing. If the metaphor of the early digital era was files and folders, the model of tomorrow's information systems will be maps and networks.

Internal networks can also be a powerful platform for innovation. In 2011, CEMEX created an internal social network for its global workforce, which consisted of more than 47,000 employees. One of the big payoffs of the project was product creation. The 600 employees who were dedicated to developing new cement products were able to collaborate on the network, and, just four months after the project's conception, CEMEX launched the first global brand of its new Ready Mix product.

Nevertheless, culture can present a challenge for global organizations. Some cultures, particularly those in Asia, are reluctant to comment on or share ideas in public forums where hierarchy is not clearly defined or there is a chance of social embarrassment. Other behavioral issues like domination or bullying are additional risks faced by organizations as they take the necessary, but sensitive, step of building more humanity into their communication platforms.

How could senior management adapt their communications behavior to encourage the adoption of an enterprise network?

Enterprise Virtualization

Enterprise virtualization refers to the separation of the infrastructure and business process of a company from its management and control.

WHAT IF YOUR entire company were virtual?

Companies are both supported and limited by their infrastructure. A computer server might allow you to run applications for your customers, but if there is a sudden spike in usage, you once had to rush to install more physical machines.

Eventually IT teams realized that a lot of the physical technologies they managed could in fact be virtualized. In the late 80s, people figured out how to run DOS programs on a Unix workstation. More recently, everything from desktops to servers, storage devices, data centers and even networks can be deployed as a "virtual instance". This means that not only can existing hardware be utilized more efficiently, but in many cases, entirely commoditized.

Now you can manage your entire IT environment with software rather than screwdrivers and replacement cables. In the 21st century, software rather than hardware is where the real value of a company lies. And that's good news, because you can adapt, program and automate it.

What happened in IT is also happening to other parts of the business as well. Few technology companies actually make technology anymore. They look after the design and marketing decisions, but they delegate manufacturing to a global network of component suppliers, assemblers and distribution channels. As demand increases, they can add capacity just like their IT team might add another virtual server.

Scalability is the key. Your business is only as strong as its least scalable component. In the virtualized enterprise, managers can activate internal resources or ecosystem partners in real time without worrying about their physical location, human staffing, or contractual status. They are, for all intents and purposes, part of the extended business the moment they are needed. In the future, companies will seek to virtualize any and all business processes that need to scale rapidly with demand – whether it be distribution, computation, marketing, production or even customer support.

What parts of your business are the most vulnerable to sudden spikes in demand?

Exascale

Exascale computing is the next big leap in computer engineering, with performance in the range of 1018 operations per second.

WHAT IF OUR computers were a thousand times faster than they are today?

You might model millions of blood cells as they move through an artery, and help a heart surgeon as he plans a procedure. You could do a better job at climate modeling and predicting extreme weather conditions. You could advance our knowledge of genomics, and even build a simulation of an actual human brain. All of these things are possible with enough processing time today – but with exascale computing the insights would be fast enough to be truly actionable.

The best way to put exascale into perspective is to compare it the human brain. According to Horst Simon, Deputy Director at the Lawrence Berkeley National Laboratory, at the moment our best computers are equivalent to about 4.5% of a human brain, running at 1/83 real time speed. At one exaflop, real time processing will potentially have the same processing power of the human brain at neural level.

The world's fastest computer today is the Tianhe-2 in Guangzhou, China, with more than 3 million cores. In contrast, an exascale machine may have hundreds of millions of cores, or even a billion. The real problem at that scale is power. Power is the primary design constraint in the future of super computers.

Another problem is making sense of the data. It is estimated that our brains can only receive information from the external world at gigabit rather than exascale rates – anything more would be overwhelming. Akira Kageyama and Tomoki Yamada from Kobe University in Japan suggested that one approach would be to use "bullet time", the filming technique popularized by movies like The Matrix. The bullet time effect slows down time while the camera angle changes as if it were flying around the action at normal speed. Kageyama and Yamada believe that we might be able to surround the simulated action with thousands, or even millions of virtual cameras that all record the action as it occurs – allowing users to "fly" through the simulation by switching from one camera angle to the next.

In what parts of your industry is the lack of super fast computation a barrier to further innovation?

Experience Design

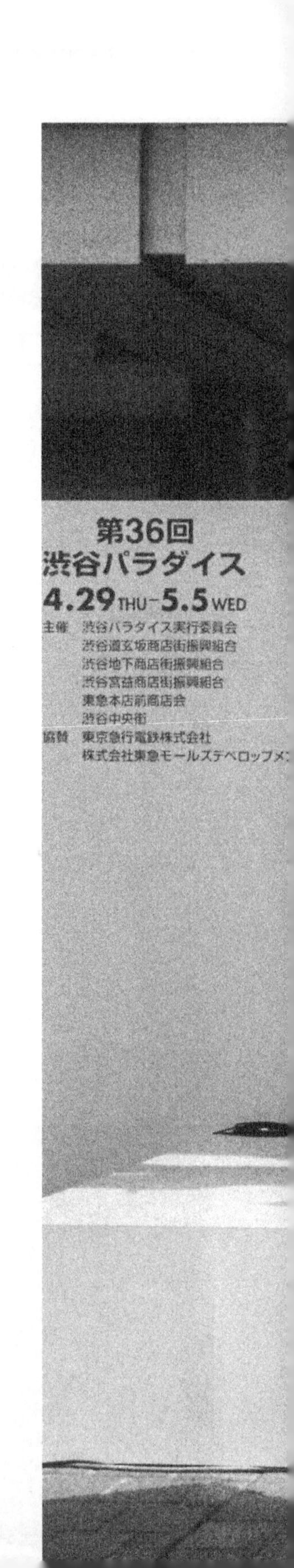

エンタテイメントシティ し
SHIBUYA PAR
きうきミニライブ

Experience design is the art of programming interactions between a brand and its customers.

WHAT IF THE way people felt about your brand were more important than what your brand produced?

Product innovation has a short shelf life. A new car design, device format or packaging concept will only remain unique for a moment before it is imitated, integrated and commoditized by a global marketplace of watchful competitors. There is one thing however, that is much harder to copy and replicate: customer experience.

Louis Cheskin was one of the original experiential marketing researchers. In the 60s, he realized that most people make unconscious assessments of products and services not only based on a logical assessment of the item itself, but on what he called "sensation transference", or the information directly provided by the physical senses. For example, he worked with McDonalds to show that their burger stands designed for walk-up service were ill-suited for families and women with children. He demonstrated that tables, chairs and partial walls created a sense of comfort, which led eventually to creation of the family restaurant format.

It may seem strange, but in the digital age, sensory design is more important than ever. Great experience designers go beyond features and benefits and focus on intangible qualities like what it felt like to buy the product, take it out of its box, and turn it on for the first time. And critically, in the 21st century, great design is also about leveraging data to achieve higher levels of personalization and customer delight.

Disney for example, has invested over $1 billion into a wearable technology called MyMagic+ at its Disney World resorts. The bands store tickets, hotel keys and payment information. They are presented to you in a personally inscribed gift box, and once attached to your wrist, help to orchestrate the total Disney experience. The bands not only allow you to move seamlessly through the park, but in the future they will also help characters great you by name or wish you a happy anniversary.

The magic behind next generation experience design is not the data you use but the logic of how all the interactions come together. As Dave Morin, CEO of mobile social-networking company Path, puts it, "AI is the new UI."

If you were to map the customer journey at your business, what would be the best and worst moments in the experience?

Fab Theft
Freelancers

Fab Theft

САНКТ-ПЕТЕРБУРГ
МУЗЕЙ
ВОСКОВЫХ
ФИГУР
2 ЭТАЖ

Fab theft is the use of digital scanning and 3D printing to replicate copyrighted designs or patented objects.

WHAT IF THE best way to steal something in the future is just to print it?

Consumer 3D printing technology is rapidly evolving to the point where it's no longer just a curiosity for hobbyists. Better quality scanners combined with higher resolution printers and more durable feedstock, are for the first time making both the commercialization and the criminalization of 3D printed objects viable.

We are on the brink of an impending clash between copyright and the future. In February 2013, HBO sent a cease-and-desist letter to Fernando Sosa asking him to stop selling his 3D printed iPhone dock that he based on the Iron Throne from TV series, Game of Thrones. Even though Sosa had designed the 3D model himself, HBO claimed it owned the rights to all images that appeared on the show.

The toy manufacturer Hasbro adopted a rather different strategy. Noticing the unexpected growth in a subculture of consumers modifying My Little Pony toys, they announced a partnership with the 3D printing platform Shapeways to provide licenses for consumers wanting to create their own fan art.

There are some similarities between fabrication theft, and the early days of the Internet, when fans who built websites dedicated to their favorite Disney or Stars Wars character, were rewarded with stern letters from legal departments. A decade on, not only have the legal threats ended, but marketing departments now actively entreat consumers to remix, blog and share their content.

The most likely fabrication thieves are less likely to be households as commercial 3D printers with the scale and quality to rapidly copy and re-distribute designer and high value items.

One way for brands to circumvent these issues may be to make their designs available to print in a secured fashion. Authentise for example, recently launched 3D Design Stream, an API for 3D marketplaces that, like Netflix or Spotify, streams designs directly to buyers' printers for a single use.

How will streaming rather than selling physical products change the business of manufacturing?

Freelancers

..can't buy
me love!

Freelancers are part-time workers who use the Web to find work and collaborate on projects.

WHAT IF THE best job were not having one at all?

There is a perception that the growth in the freelance economy is mainly due to workers from emerging markets seeking routine tasks for low wages. Several years ago, that may have been the case. These days, new technologies have made finding work easy, and simplified the process of collaborating on complex projects and collecting money.

According to a 2014 study by the Freelancers Union and Elance-oDesk, the US workforce is now 34% freelance. This is equal to roughly 53 million people, up from the 42.6 million counted 10 years ago by the U.S. General Accounting Office. What is interesting about the current rate, is that 14.3 million of these people actually have full-time jobs and are doing independent work in their spare time. Unsurprisingly, the study found that the demographic most open to freelancing were Millennials.

There are now a large number of startups connecting freelancers with work. UberX and Lyft pay car owners to drive. Elance-oDesk, Guru and RentACoder are marketplaces for programmers and developers. Taskrabbit is for odd jobs and errands. 99designs focuses on design, while Fiverr reverses the model and allows freelancers to compete with descriptions of their ideal job.

Freelancers present an integration challenge to companies. Contingent workers are an essential part of a company's ability to scale resources in growth periods, but can be difficult to combine with existing teams and connect to a company's internal networks. You can expect the management of a "blended workforce" to be a regular topic of conversation for HR professionals for some time to come.

It is not yet clear how the shift towards contingent work will impact the economy and the way in which people spend money. The next generation has already demonstrated a reluctance to follow the Baby Boomers and embrace the kind of high stress careers required to service home loans, expensive cars and other material assets. According to a report by Edmunds.com, the number of cars in the US purchased by the 18-to-34 demographic fell almost 30% between 2007 and 2011.

How might a team of freelancers approach your next project differently?

Gamification
Genome Editing
Growth Hacking

Gamification

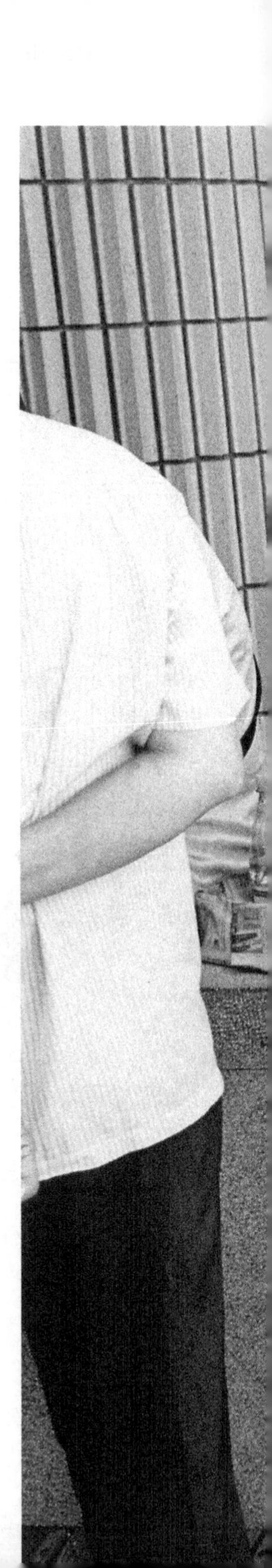

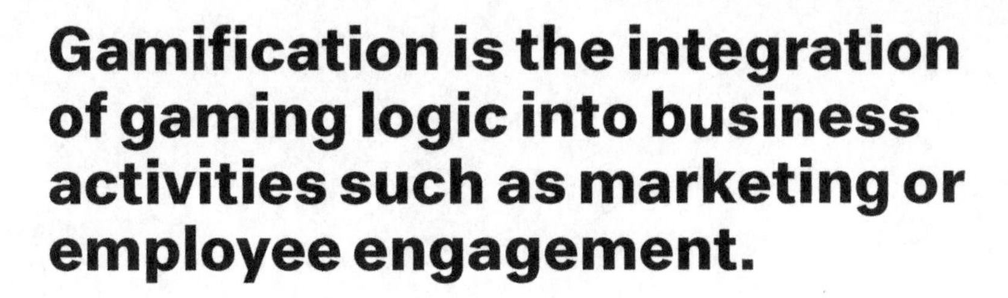

Gamification is the integration of gaming logic into business activities such as marketing or employee engagement.

WHAT IF THE best way to get people to do what you want is to convince them they are playing a game?

If you have ever wondered why online games or slot machines are so addictive, it's because they utilize what psychologists call operant conditioning. Like the infamous "Skinner Box" that teaches animals to perform certain actions in response to specific stimuli, games reward repetitive actions by using an operant schedule of reinforcement. Charmingly, those same tricks are now being used in business.

Familiar game elements such as scoring, badges and leaderboards are starting to be used as strategies to motivate employees or customers to behave in particular ways. The German software company SAP for example, uses points to reward top contributors on its SAP Community Network. Rankings are visible on a global leaderboard, used in employee performance reviews and assessed for eligibility in project teams. Some employees even use their SCN ranking on their resumes.

One of the reasons why gamification works is because it is based on continuous feedback. Rypple, which was later bought by Salesforce.com and rebranded as Work.com, used real time thanks and recognition to disrupt the rather tired idea of performance reviews. The software was subsequently adopted by both Spotify and Facebook as it resonated with younger employees whose digital habits had created an expectation of regular feedback and interaction.

Speaking to Alicia Tillman, VP Marketing and Business Services from American Express Business Travel, I learned about their initiative to use gaming to deal with the problem of policy compliance. Travel, for most companies, is either the second or third largest spend category after human capital or technology costs. Amex has created a platform called GoTime, which offers missions based on different elements of a company's travel policy. Points unlock badges and allow employees to move up a leaderboard. During the program pilot, one Amex client saw a 7% increase in bookings with preferred airlines and a 12% jump in advance purchase pricing breaks.

What is an unloved process at your business that might be improved by introducing more feedback and competition?

Genome Editing

Genome editing allows DNA to be inserted, replaced, or removed from a genome using molecular scissors.

WHAT IF WE could design smarter, healthier, better-looking people?

The human body is made up of about 20 trillion living cells, each containing 23,000 genes – collectively known as the genome. You can think of the genome as an instruction book for making and running the hardware and software platform that is a human being.

Needless to say, if you want to start playing with the genome, you first need to know how to edit it. Xingxu Huang, a geneticist at Nanjing University in China, and his colleagues recently engineered twin monkeys with two targeted mutations using the CRISPR/Cas9 system. CRISPR stands for clustered regularly interspaced short palindromic repeats – in other words patterns of molecular elements that act like scissors and cut strands of DNA when they encounter a predatory virus. The big breakthrough was when scientists figured out how to use this bacterial self-defense mechanism as a very precise gene-editing tool.

Genome editing has the potential to revolutionize gene therapy. Timothy Ray Brown was diagnosed HIV positive in 1995. He underwent a stem cell transplant in 2007 to treat leukemia. The donor was among a tiny percentage of people naturally resistant to HIV infection because they lack CCR5, a protein on the surface of immune cells that the virus uses as an entry portal. After the transplant, Brown was cured of his pre-existing HIV infection, highlighting the potential role of genetically engineered stem cells in a general cure.

While genomic editing will eventually be possible on a larger scale, it will be first more easily achievable on babies. Prenatal diagnosis or testing for diseases on an unborn baby is already a common procedure. With IVF babies, advanced screening techniques allow doctors to examine the genetic data for abnormalities, allowing for the identification of the best embryo to place in the uterus. Patients are already using this technology for sex selection. In the near future, it will not be difficult to also probe for other genetic qualities or pre-dispositions.

We will soon face the challenges of a world where wealthy individuals will be able modify their DNA to eliminate disease, provide their children with extra abilities or make themselves stronger or more attractive.

What happens when the new class system is not only economic but genetic?

Growth Hacking

Loschwassereinspeisung
INTEC
INTEC
INTEC

Growth hacking is a marketing strategy that replaces elements of traditional marketing with ideas that can be tested, tracked, and replicated at scale.

WHAT IF YOUR marketing budget were zero?

In a blog post that ended up defining an entire movement, Sean Ellis, entrepreneur and founder of Qualaroo, wrote, "A growth hacker is a person whose true north is growth." He argued that for startups, hiring a VP of marketing who could write an elegant plan was not as valuable as someone, potentially with a sales or engineering background, who took a data-driven approach to gaining customers and driving revenue. Strategy, in his words, was for nothing if a "scalable, repeatable and sustainable way to grow the business" had not been found.

Growth hackers are the ultimate marketing agnostics. They don't care what the tool is – social media, viral marketing, content, or clever landing pages – so long as it achieves measureable results. Creativity comes from constraint, but that's not the only reason why growth hacking techniques are important. A great Web business should be scalable by design. If you need billboard and television ads to gain traffic, as was the case in the first dotcom boom, then you have a serious problem.

In the early 2000s I was working for the Murdoch family in their Australian newspaper business. Their classified business was built on the back of the recruitment industry, but I was convinced that to survive the transition to digital, they needed an entirely new business model. I went to Silicon Valley to meet with a brand new startup called LinkedIn to see if I could help broker a deal. I'll never forget meeting with the company's founder, Reid Hoffman.

The office was a small, non-descript building in Mountain View with a handful of employees. I asked Reid why he believed LinkedIn was going to be so successful. He said nothing for a moment, and then, with a knowing smile said that what made LinkedIn special was a quality that he believed essential to any online venture: a built-in distribution model. The mark of a great digital business, he explained, is one that markets itself.

If your marketing budget were cut tomorrow, what would be the most effective things you could do to continue growing your business?

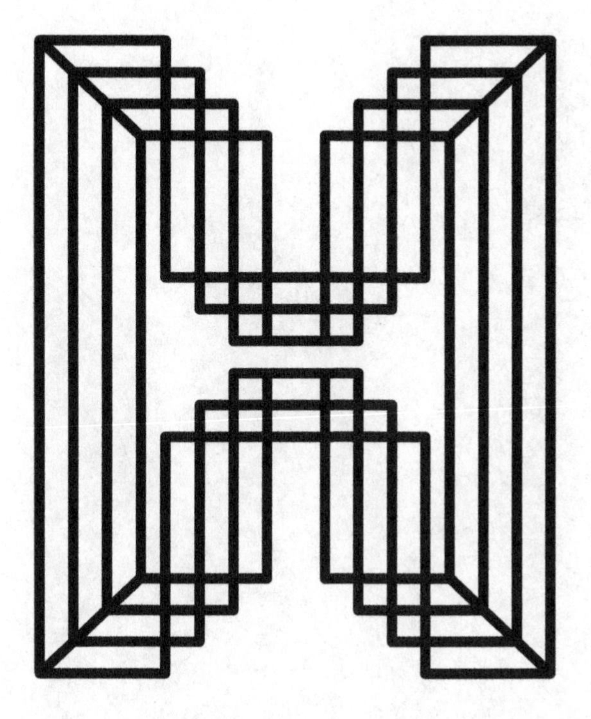

Hacktivist
High Frequency Trading
Holocracy

Hacktivist

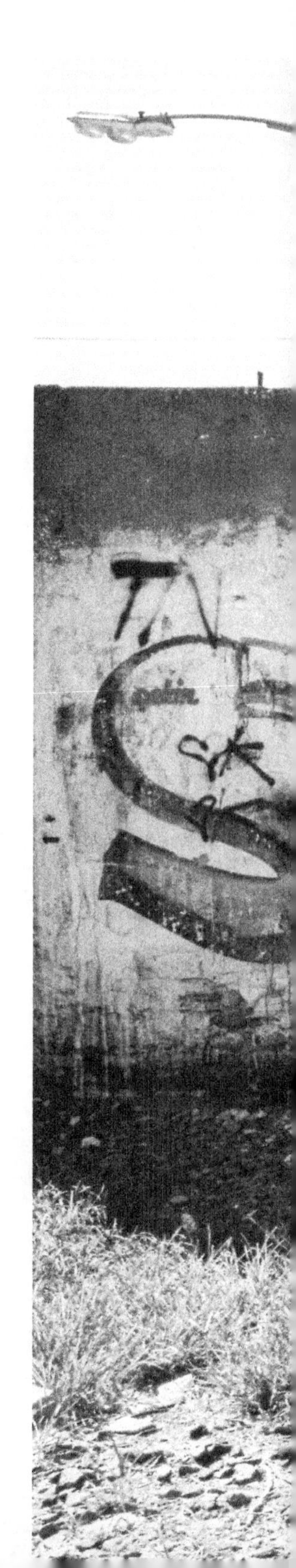

BTC

Hacktivists subvert computer systems to further social or political aims.

WHAT IF THE future of protest were online?

Hacktivism is subversive in a way that activism no longer is. Activism has evolved from grass root movements to political action committees. Activist groups like PETA (People for the Ethical Treatment of Animals), China Labor Watch, or Greenpeace are large, well funded entities that have largely swapped radical action for social credibility. And whereas activism seeks awareness of issues through traditional channels, hactivism embraces the disruptive tools of the 21st century.

The poster child of modern hactivism is the group Anonymous, which emerged from the digital primordial soup that is the 4chan message board. 4chan originally started as an "imageboard" dedicated to Japanese anime culture. The name Anonymous came from early members of the group posting on the board under that name, creating a long-running joke that a single person named Anonymous was simply talking to themself. It was an appropriate metaphor, because the group soon became a kind of distributed intelligence with an emergent leadership.

Anonymous' first operation was Operation Chanology, which crippled the Church of Scientology with a distributed-denial-of-service (DDoS) attack. Later targets of the group included child pornography sites, Paypal and Mastercard. The attention these attacks garnered increased the number of members or "anons" who, when gathered in person, sport stylized Guy Fawkes masks.

Other hacktivist organizations seek to effect change through radical exposure. WikiLeaks, represented by Julian Assange, publishes and distributes secret information from anonymous sources. It rose to prominence in 2010 with a release of U.S. State department diplomatic cables.

Acts of hacktivism present complex ethical puzzles. Is Edward Snowden a hero for revealing the Orwellian overreach of the NSA, or a traitor for endangering the agents that target terrorism? Was the overzealous prosecution of Aaron Swartz over his uploading of academic journals a fair response to a valid crime, or was the real issue the failure of his university MIT to protect him?

Where as a society should we draw the line between freedom of speech and terrorism in the digital world?

High Frequency Trading

High Frequency Trading is the use of sophisticated technology and computer algorithms to rapidly trade securities.

WHAT IF THE world's best traders were computers?

We had our first glimpse of the destructive power of automated trading in October 1987, when the exponential use of computer programs that automatically sold index futures if the market declined, led to mass panic.

There has always been an advantage for traders in being fast. In recent years, the quest for low latency has seen traders co-locate in exchanges next to trading platforms. Each additional mile of cable adds an estimated eight microseconds of lost time. When a group of scientists at the OPERA project mistakenly observed neutrinos traveling faster than the speed of light, some believed that a trading system could be built using a particle accelerator with the ridiculous outcome that a trade might arrive before you sent it.

High-frequency trading algorithms incorporate order flow information from fragmented markets and execute trades across these markets and trading platforms. Liquidity is spread among both traditional equity exchanges as well as alternate trading platforms known as "dark pools" that allow the anonymous trade of large blocks. Traders use high-speed connections and advanced algorithms to exploit inefficiencies in this new market structure. Their aim is to identify patterns in other participants' order submissions and trade executions that they can use to their advantage.

The problem arises because if you have access to the fastest connection and are the first to receive information about order flow and trades, you might be able to detect a large order from an institutional investor and then instantaneously trade ahead of this order, driving up the prices artificially and eventually selling your accumulated stock inventory to the institution at a higher price. High-frequency trading firms dispute this characterization and say that they are just providing market-making services more efficiently.

There is no doubt however that financial markets are now being transformed by advanced computation as well as automation. As trading starts to rely on greater levels of technology, it ultimately raises the question of whether we are reaching a point of such great complexity that we will no longer be able to understand, let alone control the results.

What other markets would be radically transformed with faster processing speeds?

Holocracy

Holacracy is a self-governing system where employees have different roles within overlapping circles, but no specific area of responsibility or reporting.

WHAT IF YOUR company had no bosses or job titles?

At its quarterly meeting at the end of 2013, Tony Hsieh, the founder of Zappos, introduced a radical new idea. The company had decided to eliminate job titles in favor of Holacracy, a model proposed by entrepreneur Brian Robertson as a way of replacing traditional management with the principle of democratic decision-making.

Derived from the Greek word holon, which means a whole that's part of a greater whole, the Holacratic model substitutes top-down management with a flatter holarchy that distributes power more evenly. Companies are made up of different circles, and employees can have a number of roles within those circles. In Holacracy, each circle must meet the purpose as defined by its higher circle. Purposes can be as wide ranging as managing reputation, delighting customers, or handling operations. Although there are no managers, there are people known as "lead links" who have the ability to assign employees to roles or remove them.

Hsieh embraced Holacracy based on his belief in the importance of agility. Writing on the company website, he observed that "every time the size of a city doubles, innovation or productivity per resident increases by 15%. But when companies get bigger, innovation or productivity per employee generally goes down." His goal was to structure Zappos more like a city and less like a bureaucracy, allowing people to be self-organizing.

Medium, the company started by Evan Williams, former founder of Twitter, has also implemented the idea. Some of the key tenets adopted by Medium include maximum autonomy, the elimination of people management and organic expansion, as well as making policies and decision-makers explicit.

Managing compensation is an issue complicated by distributed management models. At Valve, a gaming company that also has a boss-free environment, a peer-assessed stack ranking system is used to determine pay levels, based on your perceived contribution to the business. Perhaps the real value of Holacracy lies in its critique of modern management and its role as a catalyst to upgrade the operating system of the 21st century company.

How much time do the employees of your business spend on advancing their position as opposed to creating actual value?

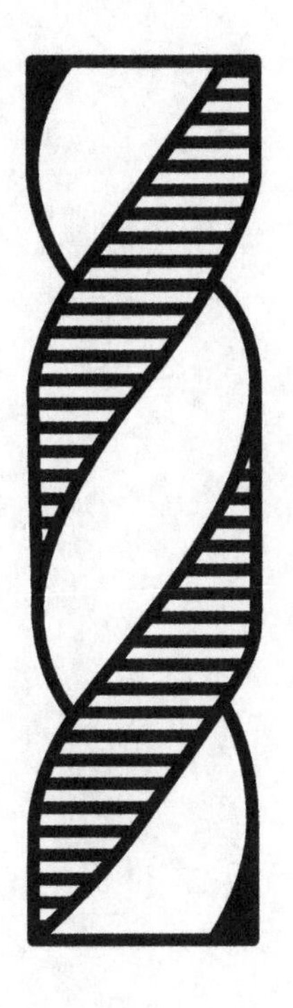

Inbound Marketing
Innovation Ecosystem
Internet of Things

Inbound Marketing

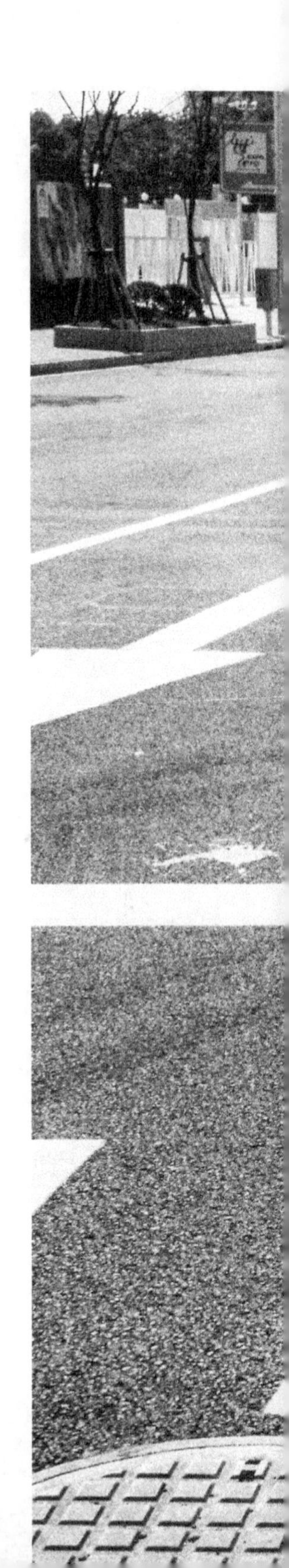

Inbound marketing is the strategic use of content, data and decision science to attract and convert potential customers.

WHAT IF YOUR content said more about your brand than your advertising?

Inbound marketing is a concept coined by the founders of Hubspot, a marketing automation firm. Looking at the marketing industry, they saw that brands were using their websites and social media channels like megaphones – broadcasting to anyone who would listen – rather than as a hub to connect directly with individual consumers.

Here's what made that idea dangerous. Rather than buying advertising to drive traffic, Hubspot suggested that brands should instead invest in content that might attract the right kinds of customers. In order to make this strategy work, inbound marketers would need to use an automation system to capture people's information when they found the site, and then carefully nurture this database with tailored emails and personalized content. The purpose of these campaigns was to move leads through a virtual sales funnel until they could be converted into customers and advocates of the brand.

To successfully implement an inbound marketing strategy requires more than just great content. It also needs an integrated approach to both technology and organizational design. Your publishing, analytics and customer databases have to work together seamlessly. If you can link your engagement system to your system of record, you can also start to understand not just what drives sales, but how much value each of your marketing activates contributes to a successful outcome.

Most importantly, inbound marketing challenges the organizational divide between marketing and sales. In the 21st century organization, these two functions are really one and the same. It is the job of marketing to attract potential leads and nurture them until they can be sold to. Then it is the job of the sales team to get them over the finish line. The key to the new model of marketing is to realize that there is a seamless path of data that connects all the behavior of your customers as they move from interest to advocacy.

To what extent do you actually influence your customer's path to purchase?

Innovation Ecosystem

John

An innovation ecosystem is a community of interacting individuals and complementary organizations that build collective value.

WHAT IF THE most valuable part of your company were not your assets, but the community you were a part of?

An ecosystem is an idea from biology. However in the 90s, researcher James F. Moore also applied the idea to business, describing the co-evolution of suppliers, producers and competitors within an economic community. In Moore's view, the development of the community tended to follow a leader, who in addition to promoting a shared vision, helped the participants align their investments.

Apple is a powerful example of how an ecosystem of content, application and accessory development is not only a source of profits, but a strategy for locking customers into your platform. Google's Android ecosystem is a different, but similarly effective, strategy. Unlike Apple's proprietary operating system, Google has created an open software platform that allows mobile device brands to innovate with their own hardware. Once again, a diverse range of participants are involved in creating overall value for the platform.

I met Sanjay Purohit, head of Infosys products, platforms and solutions at an event in Portugal. He explained to me that ecosystems must be open and allow bi-directional engagement in order to function. He gave the example of the work Infosys was doing with P&G. P&G had difficulties with the myriad of distributors in markets like India, so Infosys created Tradeedge, a platform for linking suppliers and distributors, that now allows other brands to plug-in as well.

Sometimes a more disruptive approach is required to kick-start an-ecosystem. On June 12 2014, Elon Musk announced that he was taking down the wall of Tesla patents in the lobby of their Palo Alto headquarters. He had decided to offer most of his company's patents to his rivals. He knew that alone, he would never achieve enough scale. He needed other companies to use and help enlarge Tesla's existing network of 97 charging stations, making long-distance travel more practical. Within days of his announcement, Nissan and BMW announced they were "keen on talks" to cooperate on charging networks.

Does your company just sell products to make a profit, or does it provide an ecosystem for other participants to do the same?

Internet of Things

OMRON
OMRON
OMRON

The Internet of Things is an emerging network of connected objects able to sense their environment and share information.

WHAT IF ALL the objects in the physical world could talk to each other?

Describing the future potential of the Internet of Things (IoT) is like standing at the dawn of the age of electricity, trying to accurately predict how such a revolutionary technology might transform life and industry. Similarly, when it comes to the IoT, we are perhaps also destined to overestimate the trivial (washing machines with WIFI), and underestimate the profound (self-aware supply chains).

The IoT is developing on a number of fronts. It is becoming cheaper to pack everyday objects full of sensors. Device connectivity is increasingly pushing that sensor data into the Cloud, where the falling price of computation and storage is making it possible to gain very precise insights into the physical environment and events taking place in it. This in turn is leading to new kinds of real time applications that have the potential to make the physical world completely interactive. The future is not the World Wide Web; it is the Web Wide World.

One of the key areas to watch will be manufacturing, where the IoT will drive the digitization of the entire supply chain. Manufacturers will hire world-class mathematicians to design algorithms capable of handling the complexity of millions of moving pieces which are all aware of each other. Raw materials rolling into a car plant will know they are tagged for an individual customer, while retailers will ship products to customers even before they order them. Like sharks, the supply chain of the future will stay in constant motion.

We have already reached the point where the value of an object lies less in its physicality than the code that encapsulates it. As Marc Andreesen, who created the first Web browser, put it, "software is eating the world". The merger of software and hardware is more than just convergence; it allows companies to radically redefine the purpose of objects, and by extension, entire markets.

What objects, should they suddenly become networked and digitally active, would have the most meaningful impact on your life?

Junk Patent

Junk Patent

K
O'REILLY'S PUB
ROOFTOP BAR + RESTAURANT
ATM
24 Hour
O'Reilly's
PUB
O'REILLY'S

A junk patent is a government granted protection for a design or idea that has not been fully examined or certified.

WHAT IF THE patent system were the enemy of innovation?

Wu Changshun might not strike you as a likely candidate for big ideas. And yet the recently disgraced chief of police in Tianjin is listed on 35 patents filed between 1999 and 2013, including a "multi-paneled" traffic light. Despite his apparent creativity, he is nevertheless eclipsed by the former police chief of Chongqing, Wang Lijun, now infamous for his role in the Bo Xilai scandal. Wang filed 211 patents in 2011 alone, one of which was for a rather fetching set of red raincoats for female patrol officers in his district.

You have by now probably figured out the scam. Both former police chiefs were party to a new type of patent abuse in China. Here is how it goes: officials with power over procurement in an area acquire the patents for particular products, and then award a contract to a local company obliging it to pay official royalties for the use of that patent.

No patent law existed in China before 1985. That's changed, but importantly China's patent system includes the issuance of IP assets, including utility models and design rights, often without proper examination, leading to "junk patents". The number of patent applications filed in China climbed 26.3% in 2013 to 825,000 applications filed. China overtook the US and Japan for the first time in 2011.

Part of the reason for the patent boom is the Chinese government's ecosystem of incentives. These subsidies have been extended internationally, such that the government pays for foreign filing fees. This has led to a deluge of dubious applications lodged in a variety of global markets. Of course, not all applications filed in China are junk. According to the World Intellectual Property Organization, the top three companies filing applications include ZTE and Huawei for legitimate communications inventions.

However, Chinese patent trolls are a real threat to global players. In 2014, Apple lost a patent-infringement case to Zhizhen Internet Technology, which claimed that Apple's virtual assistant Siri infringed its speech recognition patent. It wasn't the first time. In 2008, Apple was sued over the iPod. Nor are they alone. Schneider Electric lost a $48 million case over a utility model patent, while Samsung lost a similar case to the tune of $7.4 million.

Is the protection of invention an incentive to innovation or a barrier to global development?

K-Scale

K-Scale

立入禁止

K-scale (Korea scale) is a checklist for diagnosing and evaluating Internet addiction.

WHAT IF THE Internet were making us crazy?

The first time I visited South Korea in the early 2000s was like stepping off a subway into the future. Long before the iPhone arrived, I saw kids with sleek mobile devices equipped with live TV feeds, darkly lit Internet cafes packed with state of the art online gaming rigs, and a country obsessed with Cyworld, an online social network that had a 90% penetration rate among people in their 20s. All that innovation was the result of government policies to invest heavily in broadband infrastructure. So it was no surprise that South Korea was also the first country in the world to experience the dark side of connectivity – Internet addiction.

In 2010, a South Korean couple was arrested over the death of their 2-month-old daughter, who starved after her parents left her at home on several occasions for six or more hours at a time. They apparently were busy feeding their virtual baby in an online game. In another incident, a 22-year-old South Korean died of heart failure after playing StarCraft for 50 hours nonstop.

Recognizing the serious risk to mental health, psychologists in South Korea created the K-Scale for Internet addiction. The diagnostic survey includes questions on a range of topics, including length of daily use and interference with school work or work, whether the individual has fantasies about being online when not logged in or has tried to restrict their usage and failed, and whether being unable to access the Internet causes depression, anger, or unusual changes in mood.

If all that sounds like a checklist for alcoholism, you wouldn't be far off. South Korea's parliament is currently considering a law that would classify online gaming as an addiction alongside gambling, drugs and alcohol. China and Japan have had similar issues with prolonged exposure to mobile and online platforms. The problem, in a world where the Web is so embedded in daily life, is how do you help people recovering from Internet addiction when total abstinence is not a practical option?

Does technology encourage
or inhibit human connection?

Life Extension

Life Extension

著名
製龜苓膏
涼

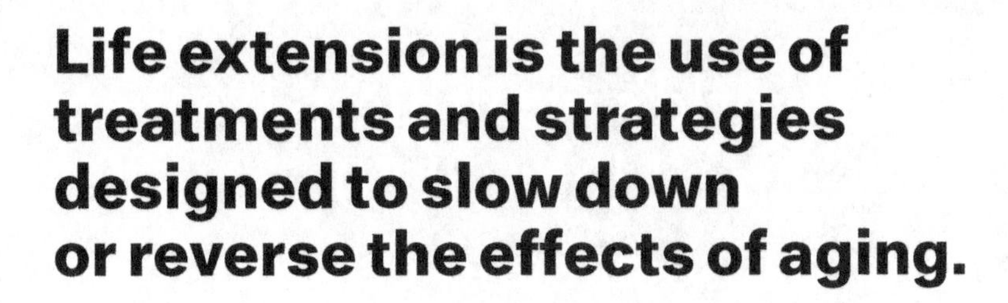

Life extension is the use of treatments and strategies designed to slow down or reverse the effects of aging.

WHAT IF WE could live a lot longer than we should?

Currently the oldest verified human life span was that of Jeanne Calment. She was born in France in 1875 and died in 1997. The scale of change she must have witnessed over her lifetime is hard to fathom. She described Vincent van Gogh as "dirty, badly dressed and disagreeable" – she had met him at the age of 13 after he had come to her father's shop to buy paints.

In 2008, I was invited to give a talk at Tim O'Reilly's Emerging Technology Conference. It was a wonderful meeting of alpha geeks and tech luminaries like Jeff Bezos. I asked the conference chair, Brady Forrest, what he thought the next big wave of disruption would be. He pointed at Bezos and said that it was not well known that groups of wealthy tech billionaires were secretly investing millions into life extension labs. At some point we would notice that these guys weren't actually dying when they should, and the secret would be out.

I didn't entirely believe him at the time, but it turns out he was probably right. In 2013, Google funded the health startup Calico, a research facility focused on longevity, with Arthur Levinson, chairman of Genetech, as the CEO. Calico has since partnered with AbbVie, a biopharmaceutical company with a view to potentially co-investing up to $1.5 billion in the project. Russian tech billionaire Dmitry Itskov has also funded his own longevity center, while a number of wealthy individuals have signed up for the Alcor Life Extension Foundation's cryogenic suspension service to preserve their bodies upon death.

One of the driving ideas behind current life extension research is that aging is not necessarily inevitable, and that the human body is more like a machine. From this perspective, genetics is the software that drives our physical hardware, and in the future, gene therapy, 3D printing using stem cells, cloned organs, and the use of medical nanobots may allow us to not only patch the code, but repair broken components. The technology is still elusive. A nanobot for example, would need to be tiny enough to squeeze through the narrowest capillaries in the human body in order to work like an artificial mechanical white cell, seeking out unwanted bacteria, viruses, or fungi in the bloodstream. But where there is money, progress will be made.

If the world's wealthiest 1% were able to live an additional 50 years, what potential social and economic impacts might this have on the other 99%?

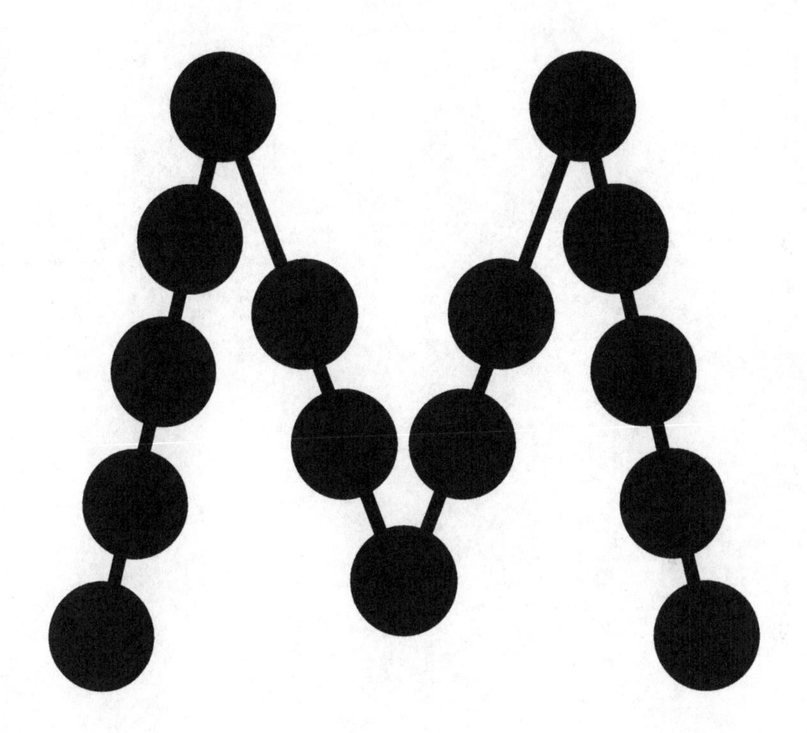

Maker Culture
Mesh Network
Metadata
Microsat
Mobile First

Maker Culture

Todoobra
ESTRUCTURAS TUBULARES
BGH

The Maker Movement is a community of hobby inventors and builders, enabled by technologies like 3D printing and computer-aided design.

WHAT IF TECHNOLOGY were less like mass production and more like craft?

The TechShop in San Jose reminded me of my Uncle's garden shed. My Uncle had been a charming, doddering old man whose private den in England was full of random tools, broken machine parts and cryptic drawings. It was only many years later that I found out he had worked on the Exocet missiles used in the Falklands War. TechShop was similarly stocked with magical devices – CNC tools, 3D printers, giant vats full of discarded electronics, and even a ring for battle robots to duke it out.

The rise of TechShop and other "hacker spaces" are part of the new Maker Culture, a reboot of the traditional arts and crafts movement combined with the technical savvy of computer hobbyists. Makers build for love and profit. Etsy, an eBay for handmade products, now has over a million sellers, while other companies like Quirky and Kickstarter make it possible for Makers to turn their ideas into commercial products.

There are similarities between the Maker Movement and the origins of the personal computer revolution, when hobbyists in their garages and groups like the Homebrew Club helped launch companies like Apple and Microsoft. But there are important differences as well. Makers are not interested in becoming the next Steve Jobs or Bill Gates. There is a sense of personal scale and individuality about some of their products that is more artisan than entrepreneurial.

Kevin Kelly, in his introduction to the Maker bible, Cool Tools, wrote: "The skills for this accelerated era lean toward the agile and decentralized. Therefore tools recommended here are aimed at small groups, decentralized communities, the do-it-yourselfer, and the self-educated."

Makers are not unique to America. In many ways the movement is analogous to the ingenious modification and invention culture that can be seen in emerging markets like India. Unlike many countries, most of the washing machines in India are built with screws rather than rivets. Manufacturers know that their customers will want to take them apart to fix them or make them better.

If you were to take a Maker approach to your next product design, what would you do differently?

Mesh Network

In a mesh network, every node relays data to the rest of the network.

WHAT IF YOU no longer needed a telecom provider to be connected?

The idea behind Firechat, an app developed by the company Open Garden, was to allow people to exchange messages when the Internet was unreliable, like at a concert, on a subway or at the beach. What they didn't expect was the sudden, exponential growth of users in war-torn Iraq. With the government seeking to block Internet access to prevent coordination and recruitment by the Islamic State, thousands of Iraqis turned to the resilient mesh networking features of Firechat to stay in communication with each other. Open Garden estimates that only 7-8% of the population needs to have the app in order for users to be connected to someone else more than 93% of the time.

Mesh networking technology has been around for a while, but for most people it is simply too slow to use as a substitute for the Internet, unless there is a pressing need such as during a crisis or emergency. After Hurricane Sandy hit communications, Brooklyn's Red Hook neighborhood developed a mesh network that rapidly grew in popularity, while Spain has a community network called Guifi with over 23,000 users.

There are two areas where you are likely to see future growth in mesh networks. The first is connected objects. Sometimes it is not necessary for a physical item to be online, so long as it can talk to other objects. Another Open Garden product is the TrackR key finder. If you lose your keys behind your chair, they're easy enough to trace, but if you drop them in the park, there is no way for your keys to ping you. With Open Garden's platform, every time another TrackR user comes within range of your lost keys, their phone will initialize and send an alert.

The second area of potential growth will be much more personal. Your watch, phone, headphones, jewelry, and even the smart fabric in your clothing may all come to be connected using mesh networks. Every device will be a hotspot, and every human being part of a ubiquitous network.

In an emergency or crisis scenario, what kind of mesh network applications would be the most useful?

Metadata

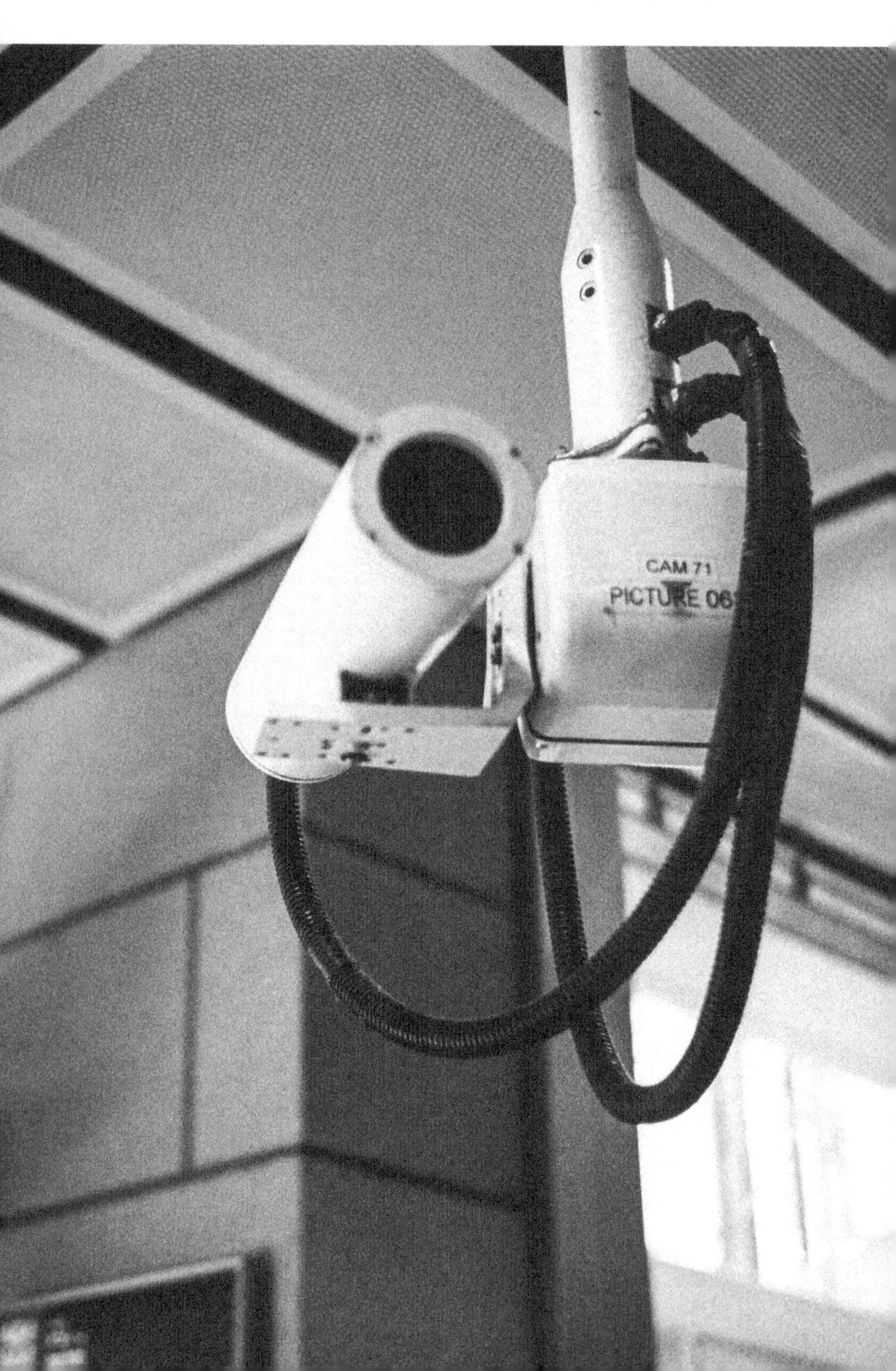
CAM 71
PICTURE 06

Metadata is data about data.

WHAT IF THERE were no such thing as privacy?

Edward Snowden's revelations about US surveillance forever changed the perception of privacy in a digital age. The NSA claimed they were just logging metadata – which sounded innocent enough; after all, they were not recording your calls, just their date and time, who you spoke to, and where you last accessed your email. And yet in most cases, even that small amount of information is enough to let authorities map a suspect's social network, or establish their physical location at a point in time.

The purpose of metadata is to assist in the discovery and classification of information. Consider a 140-character tweet. There are as many points of metadata as there are characters – including a timestamp, a location stamp, a language, an author URL, and a follower count. That information is essential for analysis, but more importantly allows the creation of applications that leverage metadata for their functionality.

The abuse of metadata comes when you start to scale and combine it with other data points. Allegedly, the NSA was also providing data to nearly two dozen US government agencies using a "Google-like" search engine, sharing intelligence on 850 billion communications made by ordinary people. Deep surveillance coupled with a great interface starts to feel a little uncomfortable, especially in a country not yet a police state.

Essentially, the debate turns on whether there is a reasonable expectation of privacy when dealing with communications metadata, and if so, whether a search warrant should be required to obtain it. Establishing a valid expectation of privacy is complicated by the fact that in this age of advanced computation, even the smallest piece of data can be easily correlated with others in a way that is highly intrusive.

To make matters worse, politicians don't always understand the complexities of the issues involved. In Australia, where a program of metadata collection is under consideration, Prime Minister Tony Abbott recently assured the public that the target "is not what you're doing on the Internet, it's the sites you're visiting… it's not the content, it's just where you have been." In safe hands, indeed.

What position will your company take on metadata? Will you be an enabler or a defender against tracking?

Microsat

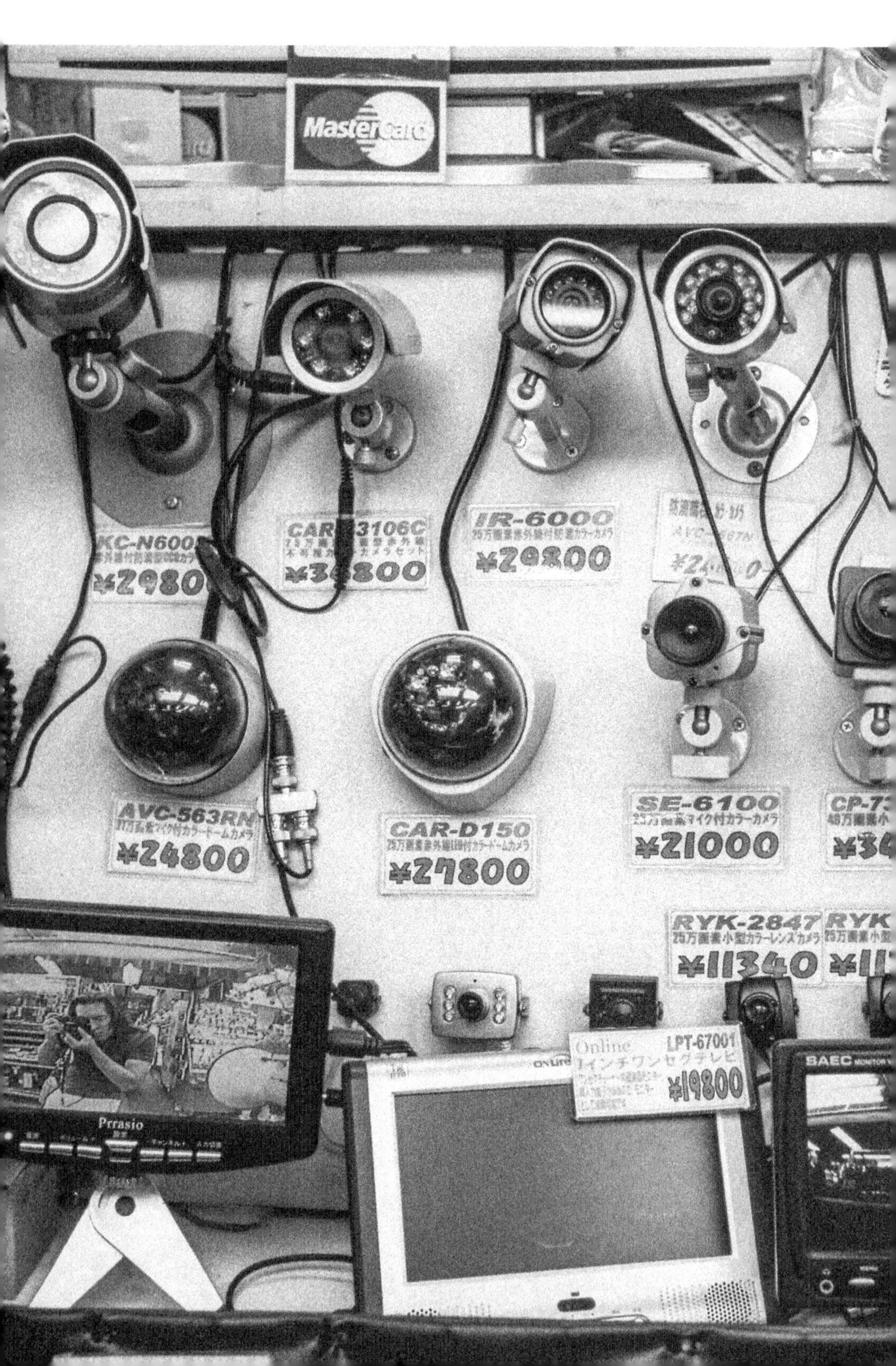
MasterCard
IR-6000
¥29800
AVC-563RN
¥24800
CAR-D150
¥27800
SE-6100
¥21000
RYK-2847
¥11340
Online
LPT-67001
¥19800
Prrasio

A microsat is a miniaturized satellite that is cheaper to design, manufacture and launch.

WHAT IF SMALL business went into space?

I remember the first Iridium satellite phones. They were sleek, sexy and the handset of choice for covert operatives, billionaires and African dictators alike. It was the late 90s and Iridium had just spent $5 billion launching its network of 66 satellites, promising a wireless service that would work anywhere on the planet. The first Iridium call was made by the then Vice President of the United States, Al Gore. Nine months later, the company filed for bankruptcy. It had defaulted on $1.5 billion in debt, and had only managed to sign up 10,000 subscribers.

If Iridium is a cautionary tale about the dangers of big budget space ventures, then the unfolding story of micro satellites is a very different one indeed. Small in size, light in weight and made of standard components, microsats can be built and even mass produced at a fraction of the cost of normal satellites. Microsats can also be launched "piggyback", using the excess capacity of larger launch vehicles. When multiple microsats work together they are known as a "satellite swarm".

One of the reasons for the growth in private sector and academic interest in satellites has been the rise of the CubeSat standard – a reference design proposed in 1999 by professors Jordi Puig-Suari of California Polytechnic State University and Bob Twiggs of Stanford University. Their idea was to enable students to design, build and operate a device with capabilities similar to that of the original Russian satellite, Sputnik. CubeSats are available in kit form and can be built using off-the-shelf electronics.

There are a growing number of microsat startups including Planet Labs, which has launched a constellation of CubeSats to create real time imagery. Skybox Imaging, a similar type of business, was bought by Google in 2014 for $500 million.

The falling price of satellites will open real time imagery and analytics to a wide variety of new applications such as insurance validation, agricultural monitoring, city planning, environmental protection, disaster response or Wall Street analysts seeking the latest data on how many customers visit retail chains. It will be a new space race – just a much smaller one.

What new types of businesses could you create with access to low cost, real time satellite imagery?

Mobile First

値札なし
100円
値札なし

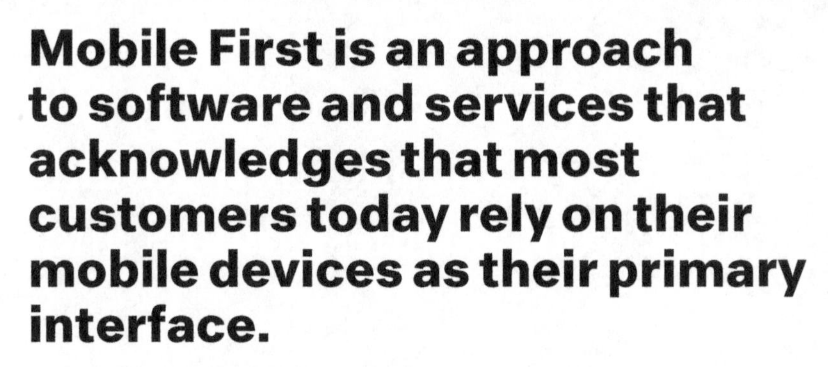

Mobile First is an approach to software and services that acknowledges that most customers today rely on their mobile devices as their primary interface.

WHAT IF ALL software were designed to work best on mobile?

When Satya Nadella took over as the new CEO of Microsoft, he wrote a very personal memo. He challenged his employees to reimagine the company for a new world, one that was mobile-first and cloud-first. Smartphones had led to the creation of what Nadella called the "dual user" – a person who engages deeply with technology both at work and at home and wants to move seamlessly between them.

The idea of catering primarily for mobile users is not as controversial an idea now as it was a few years ago – if for no other reason than the market for smartphones has exploded. The GSMA, which represents over 800 mobile operators worldwide, has forecast that smartphones will account for two out of every three mobiles globally by 2020, representing more than six billion connections. These numbers are fuelled by the falling cost of handsets, increasing demand in developing markets and the wider availability of mobile broadband.

In the US alone, Nielsen has estimated that more than 67% of US mobile users owned a smartphone at the end of 2013. That is 167 million Americans who live, work and play through their personal devices. A tantalizing market to capture, but also one that presents unique problems.

Developing for mobile devices is complex because of the wide variety of handset brands, operating systems and form factors on the market. Even more difficult is designing a complete and self-contained customer experience that doesn't require periodical maintenance or syncing with a traditional computer. In the enterprise, security and legacy systems pose their own set of challenges.

Longer term, the secret of mobile success begins with realizing that the devices themselves are irrelevant. Whether you are designing the next killer consumer app or trying to build a customer intelligence tool for your sales team, the handset itself is nothing more than a network-connected screen. What's really mobile is not the devices, but the people themselves.

If your customers could only interact with you via mobile, how would that change your business model and service design?

Nanofactory
Net Neutrality

Nanofactory

A nanofactory is a molecular assembly machine designed to manufacture products with atomic precision.

WHAT IF WE could build things one atom at a time?

When I heard that Ralph Merkle was speaking at the Singularity University, I was excited, having read about his pioneering role in creating public key cryptography. But from my perspective, his real claim to fame was when science fiction writer Neal Stephenson described him in one of my favorite books, The Diamond Age, as a founding hero of a future civilization ruled by nanotech.

In writing his book, Stephenson had been inspired by Merkle's research on nanotech, and in particular the nanofactory. In some ways, nanotechnology itself is not as dangerous and disruptive an idea as is the concept of a factory precise enough to build at the nano scale. Since the dawn of time, we have made things by casting, milling, grinding, and chipping materials. As Merkle points out, this is the equivalent of manipulating atoms in "great thundering statistical herds". A nanofactory, on the other hand, works very differently. It allows us to make products at a single atom level of perfection.

A nanofactory need not be very big. It could be small enough to sit on a desktop, and build a variety of atomically precise diamondoid products. Diamondoid is so named because of the structures you could build, which would resemble diamond in a broad sense, being strong, stiff and containing a dense network of bonds. The products themselves might range from incredibly powerful nanocomputers to medical nanorobots.

Some progress is being made in the field. Researchers at ETH Zurich, in Switzerland, recently created a "molecular assembly line" that works like an automated factory for building complex chemical substances. Just like a production line, their lab-on-a-chip system has a mobile assembly carrier that moves each product between a number of assembly stations in the form of microscopic canals, through which a solution is pumped.

Where there is reward, there is also risk. An emerging threat to this new technology is nanopollutants, particles small enough to enter your lungs or be absorbed by your skin. Nanoparticles exist in consumer products today like cosmetics and sunscreen, but in the future they may pose a significant health risk for workers handling nanofactories.

How might the rise of molecular manufacturing disrupt the basis of our industrial society?

Net Neutrality

ONE WAY
MADISON AV
AMERICA'S FASTEST,
MOST RELIABLE
INTERNET
FiOS
888.GET.FiOS verizon.com
www.verizon.com

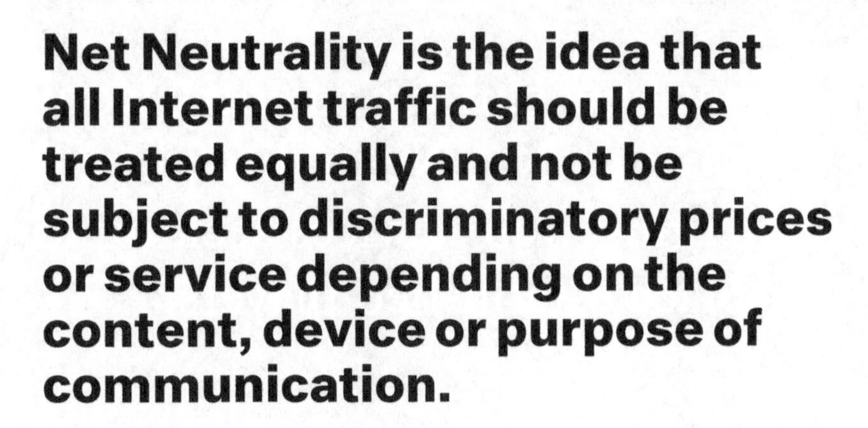

Net Neutrality is the idea that all Internet traffic should be treated equally and not be subject to discriminatory prices or service depending on the content, device or purpose of communication.

WHAT IF CABLE operators were in control of the Internet?

The spinning wheel on my screen was not a good sign. It was 4am. I was sitting in an all-night diner in Los Angeles with a pot of lukewarm coffee for company, and struggling to finish this book. The bleary-eyed staff must have thought I was a frustrated scriptwriter, and didn't even blink when I thumped the table in frustration. It was Internet Slowdown Day – a digital protest by websites like Twitter, Tumblr, Netflix, Kickstarter, Etsy and Vimeo based on the impending threat of cable companies selectively speed throttling or blocking content online.

It's a debate over what is called Net Neutrality. Content companies are becoming concerned now that cable companies have started charging websites for the faster delivery of their service. Allegedly, Comcast even slowed down Netflix's service as an inducement to get them to sign their agreement. There is the fear that if digital media businesses need to pay a heavy toll to operators in order to launch, we could end up with less innovation and a potentially very fragmented experience of the Web.

It might seem obvious that broadband carriers shouldn't be able to interfere with the Web, but in fact the FCC's ability to regulate broadband ISPs was challenged by a US federal court decision in January 2014. Basic services like phone lines are subject to common carrier laws, which stop carriers from discriminating against or refusing service to customers. However, there is another category that covers things like web hosting, which is considered an information service. The FCC made the crucial mistake back in 2002 of classifying companies like Verizon and Comcast as information service providers rather than telecom carriers. Now that decision has come back to haunt them and hinders their ability to defend the idea of an open Internet.

Curiously, on the other side of the debate are also digital libertarians who are more fearful of regulation than they are of censorship. They may dislike the behavior of cable companies, but they believe government intervention may only make things worse. As entrepreneur Peter Thiel, the first investor in Facebook, argues, "we need the US government to regulate the internet about as much as we need the EU to regulate Google".

What new barriers to entry would the loss of Net Neutrality place on emerging web services seeking to displace established players?

Open Platform

Open platforms allow third parties to integrate, add new services and extend functionality.

WHAT IF YOUR platform became more valuable the more you shared it?

You can think of platforms in a variety of ways. Google's Android or Microsoft's Windows are software platforms, while you might characterize Apple's iTunes or the Amazon cloud as business or service platforms. Alternatively, there are also platforms that work as applications, like Facebook or Twitter. A platform is less about the technology than it is something that other companies benefit from building on top of.

Platforms open themselves to the outside world through application programming interfaces or APIs. These allow third parties to integrate with the platform to add functionality, without having to ask explicit permission or request custom changes to the original source code.

Platforms benefit from network effects. The more people who use them, the more incentive there is for third parties to develop their own services. Those services extend the functionality of the platform, which in turn, attracts more users. Twitter, for instance, could have sought to control their user experience and forced people to visit Twitter.com or their official app. Instead, they allowed companies to write their own applications and interact with their data in automated ways – driving usage and the value of the platform.

Scale is another important consideration. At a certain point, the only way Amazon could continue to improve its operational efficiency was to allow other companies to use its technology and distribution services. That's why even Amazon's competitors like Apple are said to secretly use its platform. It might seem odd, but it is often in the mutual interest of competitors to co-operate in matters of scale.

Open platforms can be problematic when a lack of control leads to a fragmented experience for users. Apple was successful because it created a safe and predictable environment for its customers. AOL adopted a similar strategy in the early days of the Web. But safety can also mean stagnation.

In the future, as platforms become a more important feature of how companies interact with their customers, they may have to decide whether it is better to optimize for controlled experience or maximum scale.

To what extent are other companies in your industry reliant on your platform for survival and growth?

Permissionless Innovation
Personal Robotics
Pharmacogenomics
Pivot
Predictive Analytics
Programmatic Marketing

Permissionless Innovation

주차금지
협조에 감사드립니다.

Permissionless innovation is the idea that experimentation with new ideas, technologies and business models should be allowed by default.

WHAT IF YOU didn't have to ask before trying to break something?

Consider how many technologies would never have existed if they had gone through a formal approval process.

Vint Cerf, one of the fathers of the digital world, has argued that permissionless innovation is responsible for the economic benefits of the Internet. Can you imagine if Skype had to get a permit from every government to operate a communications service? Or if YouTube required a license to run a digital broadcast operation? Curiously, these are exactly the kinds of problems currently faced by the upstart transport coordination service, Uber.

A German court recently ruled that Uber was breaking German laws by operating without the permits, insurance and vehicle inspections required of other public drivers. Furthermore, if Uber continues to operate, the national taxi body has said it will seek to impose fines of up 250,000 euros per ride. It is not the first time Germany has tried to regulate the Web; in 2011 a German state banned companies from embedding the Facebook "Like" button on their Websites, as it was feared that these companies would then be able to track personal user data.

Adam Thierer, a senior research fellow with the Technology Policy Program at the Mercatus Center at George Mason University, argues that if government policy is guided by the fear of hypothetical worst-case scenarios, then innovation is less likely, especially since the best solutions to complex social problems tend to be organic and "bottom-up" in nature. In other words, the responsible default position for governments on new technology should be "innovation allowed".

Overzealous regulators are not the only challenges to innovation. Sometimes companies can also be responsible for stifling dangerous ideas. Departmental approvals, budget spending guidelines, job demarcation standards – all are examples of a permission-based architecture that may stand in the way of your next big idea coming to life.

What if the best way to succeed was not by doing something better, but by breaking the rules?

Personal Robotics

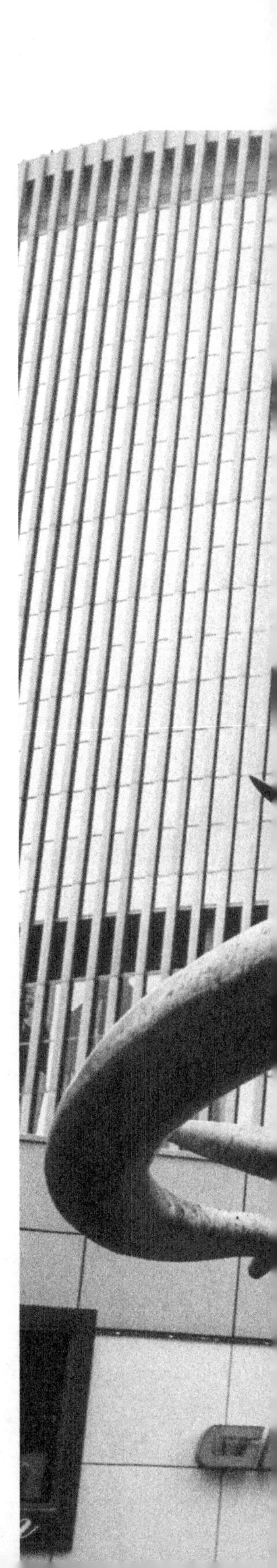

CITIZEN
松竹

A personal robot is a programmable machine designed to assist human beings.

WHAT IF YOUR children were raised by a robot?

It was 1985 and my 9th birthday. A colleague of my father's presented me with quite possibly the best present ever: an insanely sophisticated robot cat. For those of you who might remember, this was the "Petster" made by Axlon. When you stroked it, it purred. Built-in sensors allowed it to navigate a room without hitting obstacles, and if you clapped or called, it would follow the sound. It also had a rather nasty habit of waiting in the dark with its glowing green eyes and zooming toward you as you approached – which was enough to convince my grandmother that an evil spirit had possessed its microprocessor.

Despite that auspicious beginning, the home robot revolution never quite took off in the way I expected. That was my first and last robot. But things may be about to change. At the dawn of the PC era, it was inconceivable that every home would have one, let alone multiple computers. Like industrial robots now, computers then were large, expensive and purpose-built for high value tasks. Yet in some ways, the breakthroughs in computing over the last two decades have come because of the exponential growth in their use by ordinary people. Will the same be said of personal robotics?

Robots work effectively in factories because their tasks are structured for them. Home environments are more challenging – for practical as well as anthropological reasons. When he introduced his company's new line of personal robots in 2014, SoftBank CEO Masayoshi Son said that "for the first time in human history, we're giving a robot a heart, emotions". For Son, like many Japanese technologists, their frame of reference for robotics was not industrial science, but the 1960s TV show, Astroboy.

What makes the SoftBank personal robot, Pepper, interesting is that it's designed to analyze and share its interactions with people and other robots. It will observe when people laugh at a joke, or if a child responds to a particular story or book. That data is then uploaded to what Son calls the "cloud AI", allowing other robots to evolve through collective wisdom.

With its rapidly ageing population and affection for robots in popular culture, Japan has long led the world in robotics. Teaching an individual robot to interact naturally with humans is difficult. Linking potentially millions of cheap home robots and allowing them to learn in tandem, may just make all the difference.

What might be the first household app to drive the mass adoption of personal robotics?

Pharmacogenomics

DRUGS

Pharmacogenomics is the emerging field of technologies that analyze how genetics affect an individual's response to drugs.

WHAT IF MEDICINE were personalized for your unique set of genes?

Drugs don't work the same on everyone. Age, lifestyle, health and most importantly, your genes, all influence your response to medication. People may be 99.9% identical in their genetic make-up, but differences in the remaining 0.1% hold important clues about health and disease.

Pharmacogenomics is a new approach to medicine that uses a patient's biological indicators, or "biomarkers", as a guide to treatment. In simplest terms, it is a combination of pharmacology and genetics designed to predict whether a medication is likely to help or hurt you, in contrast to the "one-dose-fits-all" approach of prescribing.

So for example, if you have suffered from a recent heart attack, your physician may order a genetic test to build a profile of your response to potential drug therapies. Blood-thinners are useful in the treatment of stroke and heart attacks, but not without the risk of uncontrolled bleeding. In the future, and before you take that kind of medicine, complex algorithms able to take into account your clinical, genetic and environmental context will be set in motion in order to predict your risk of hemorrhage.

Although the cost of sequencing a single human genome has plummeted from about $10 million to a few thousand dollars in just six years, the next challenge for personalized medicine will be how to manage the massive scale of data involved. A whole genome is more than 100 gigabytes of raw data, and a million genomes will add up to more than 100 petabytes.

Nevertheless, we are likely to see numerous potential breakthroughs once we start to combine personal genomic data with medical histories and information about the environments people live in. Aggregated at scale, this may very well unlock the true mystery of the genetic underpinnings of disease.

How will the healthcare industry change when more affordable genetic sequencing shifts the focus of medicine from curing problems to preventing them?

Pivot

A pivot is a dramatic change in a company's product or business model.

WHAT IF YOU had one last chance to save your company before you ran out of money?

A surprising number of companies started off doing something very different. Twitter was the side project of a podcast directory, Groupon a platform for collective action, Android an operating system for cameras, YouTube a dating site, and Instagram a location-based check-in application.

Some see pivoting as merely a last roll of the dice before a startup runs out of money. But it is a little more complex than that. Eric Ries coined the concept in his book The Lean Startup, in which he defined a pivot as a change in strategy without a change in vision.

The key to understanding the difference between strategy and vision is to first consider the true purpose of a startup. A startup in Ries' view is not a miniature or "dollhouse" version of a big company, but a lab built to answer a simple question – is this a viable business? Or, in the words of Steve Blank, whose work on innovation inspired Ries, is there a scalable and repeatable sales process?

When a new company stumbles, it is tempting to attach blame to an individual. While firing the VP of Sales, or even the CEO may feel like a bold move at the time, it fails to address the underlying problem – the model is broken. Smart companies learn from their customer successes and failures to modify the way they price, package, or position their products. A pivot can encompass any or all of these things, or it may involve a more radical change such as technology or even what the right customer market should be.

It is not only startup companies that need to think about executing a pivot in times of crisis. Even established companies in mature industries may reach a point when disruptive technology or new patterns in customer behavior lead to a market shift. The hard thing about pivoting when you are a big business is that you don't always have the imminent threat of running out of money as an incentive. Death for big companies may be final when it arrives, but like a soprano in a Puccini opera, it can also take an entire act to happen.

How might the capabilities of your team and assets be deployed for a completely different purpose?

Predictive Analytics

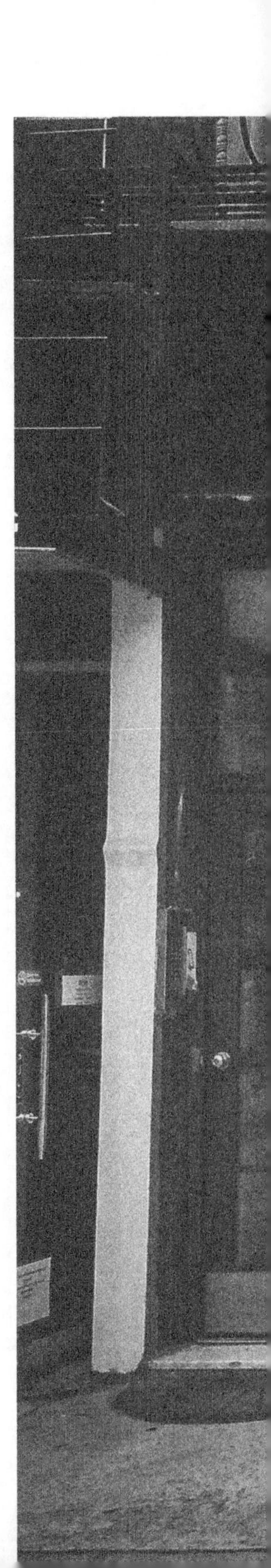

SUMMER VISIONS
&
PSYCHIC PREDICTION
BROADWAY REALTY
www.BroadwayRealty.com
AVAILABLE
THROUGH
212-577-2270
A&I Broadway Realty
WARNING
AVISO
CARDNET GROUP
ATM

Predictive analytics is the use of statistics, data modeling and machine learning to make predictions about the future.

WHAT IF YOU knew what your customers wanted before they did?

Once a week I receive a call from my bank. It's not a friendly, "how are you doing?" chat. Nor is it a clever attempt to upsell the latest financial product. It is just that I seem to persistently trip their bumbling credit card fraud prediction model, which doesn't allow for someone with a travel schedule as varied as a grumpy airline stewardess.

You can't escape predictive analytics today because in essence everything is driven by data. Predictive models exploit patterns in data to identify potential sales opportunities, recommend products to customers, spot fraud and even determine the likelihood of an equipment malfunction.

Previously, the focus of IT investment was on business intelligence platforms, which allowed managers to make decisions based on the analysis of historical data. However while business intelligence tools might tell you how fast your stock was turning or how many calls were resolved by your contact center in the past month, they couldn't predict what might happen next. More useful than last quarter's sales trends might be a set of predictive lead scores that your team can use to identify which customers are ready to close and which need more nurturing.

Bill Franks, Chief Analytics Officer for Teradata, argues in his book The Analytics Revolution that, like manufacturing in the 1800s, the field of analytics needs to go through its own Industrial Revolution. In his view, rather than crafting more highly bespoke sets of analysis, analytics now must become embedded within business processes in order to automate thousands or millions of decisions without human intervention.

Used in the right way, predictive analytics is more than just forecasting performance. Good analytics can also help leaders reframe their thinking and ask better questions. Like my own experience with the poorly calibrated fraud algorithm at my bank, a broken prediction model can tell you as much about what is wrong in your business as one that works.

How much of your business data is retrospective rather than forward looking?

Programmatic Marketing

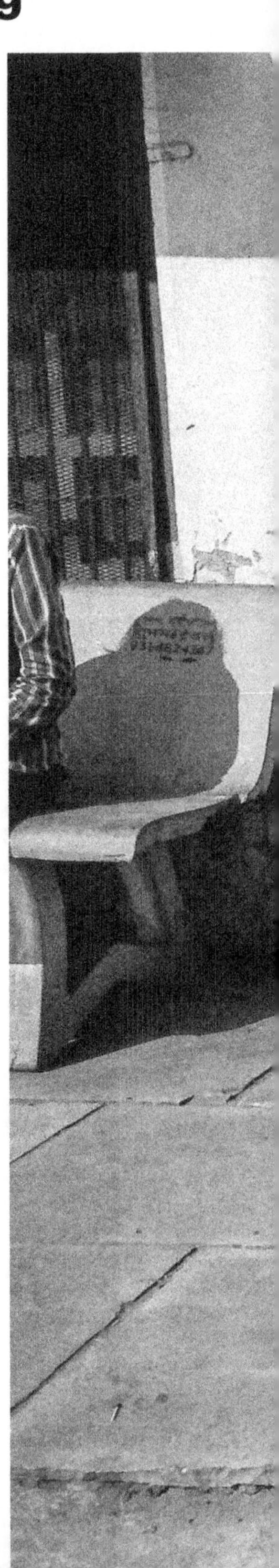

Programmatic marketing is the use of data-driven software to automate the process of buying advertising.

WHAT IF THE buying and selling of advertising were completely automated?

While brands have made considerable progress in digital marketing, the actual process of buying media has until recently remained stubbornly analogue. For a long time, advertising deals had to be arranged using an arcane mix of phone, email, insertion orders and long lunches.

Real time bidding changed everything, allowing brands to buy online impressions from publishers via an auction system, with computers making split-second decisions about what ad unit to serve. The classic application of this technology is "retargeting", whereby a consumer who views a product on a retailer site will then see an advertisement for that same product a few minutes later when they jump to another site.

The future of programmatic marketing will go beyond improved bidding systems. Publishers and platforms are starting to realize that automation is not just relevant to low value inventory, but is also the most appropriate method for pricing premium offerings as well.

Technology now allows marketers to buy people rather than placements. It doesn't eliminate human beings in the strategic planning process, but rather automates the transactional elements of buying and selling. Programmatic platforms learn through feedback cycles what kinds of advertising really work for a particular customer segment.

Eventually, traditional forms of media may also be traded algorithmically, potentially spelling the end of "prime time". For example, if you can combine set top box data with online profiles, you might discover that one of your customer segments watches a relatively unpopular TV show. This will then allow you to cost effectively target those customers using broadcast marketing at a fraction of the cost of a prime time slot.

More data does not however mean less creativity. Programmatic marketing instead encourages brands to constantly develop compelling content while aligning messaging with an individual customer's journey. If anything, programmatic technology is an invitation to finally fulfill the promise of personalized engagement.

When you hire an advertising agency in the future, how will you assess their technological and analytical capabilities?

ABCDEFGHIJKLMNOPQRSTUVWXYZ

Quantified Self
Quantum Computer
Qwerty Effect

Quantified Self

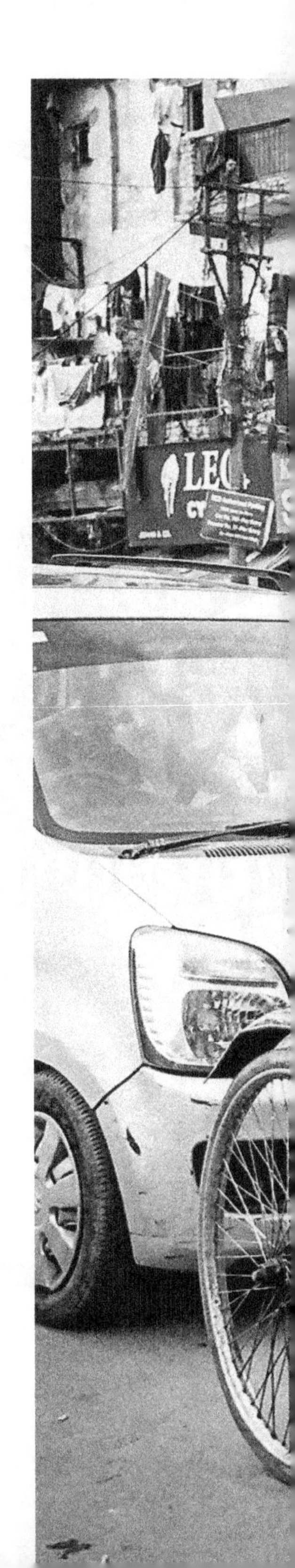

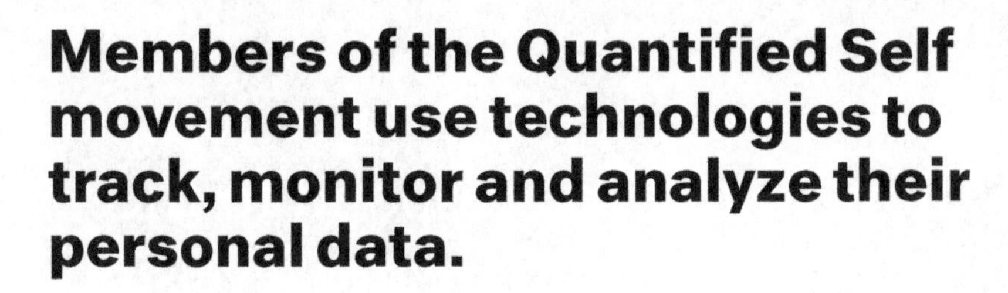

Members of the Quantified Self movement use technologies to track, monitor and analyze their personal data.

WHAT IF YOU had a record of everything you did?

The Quantified Self movement began as an online community sharing personal stories of self-tracking techniques. Now it is an entire category of consumer devices and applications, set to grow in scale as wearable computing makes the logging of personal data more accessible.

At first glance, self-tracking seems an unusual but not particularly dangerous idea. There have always been people throughout history who took an almost perverse delight in recording the minutiae of their daily existence, from the Roman philosopher Seneca to Benjamin Franklin.

Self-tracking only becomes a dangerous idea when it starts to influence behavior. With a large enough sample of data about your eating, working and exercise patterns – you might start to correlate your feelings of happiness, productivity and physical vitality with the decisions you make. Arguably, as our devices become more intuitive and responsive to our personal biometrics, they may also help us to make smarter decisions in real time.

The other side of the personal data revolution is what happens when it is shared collectively. PatientsLikeMe is an online community that allows people to discuss their personal health experiences in order to compare symptoms and the effect of treatments. Although much of the data is anecdotal rather than based on direct feeds from personal tracking devices, the forum has nevertheless had a number of successes in identifying drug interactions and new approaches to ailments.

Eventually our wearable devices will interact with platforms and professionals who will help us be proactive in every aspect of our health. We will optimize virtual simulations of our body like a game, monitoring blood sugar levels, metabolism, stress and immune systems. The challenge in the near future is how we manage expectations around the use of this data. Providing your family doctor with a real time visualization of your activities is one thing; your insurance company and HR department – well, that will be another issue altogether.

If you were an insurance company, how would you respond to the opportunity for greater levels of personal biometric data?

Quantum Computer

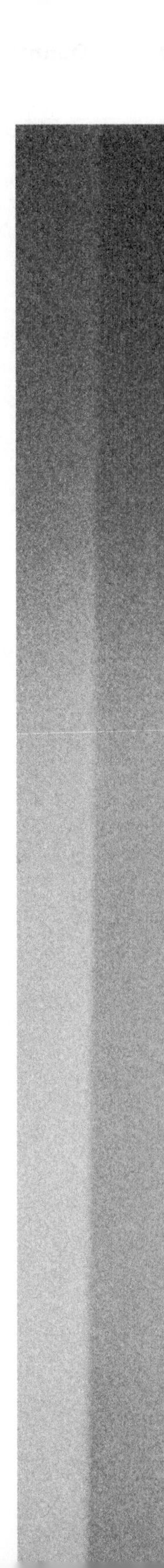

A quantum computer is a device that takes advantage of the special features of the subatomic world to perform calculations.

WHAT IF THE next big thing in computers were very, very small?

The race to build a faster computer is not particularly interesting, unless of course it involves bending conventional ideas of reality. A quantum computer is just that: an emerging architecture for computers based on the "spooky" properties of the universe at subatomic scale.

Traditional computers encode information in binary form – either as 1s or 0s. A quantum computer however, is based on qubits, which can take on several, or many values at once – think yes, no and maybe. Furthermore, quantum computers capitalize on the phenomena of "superposition" and "entanglement", where superposition is the capacity for a quantum system to be in multiple states at the same time — for example "here" and "there" or "up" and "down" – and entanglement describes the fact that quantum particles are so intimately connected that they remain linked even when separated by vast distances.

Rather than a classical computer, which has to perform each task sequentially, a quantum computer can potentially do many thousands, or millions, of computations simultaneously. This is particularly useful for optimization, the classic problem of which involves a travelling salesman who must find the shortest possible route between a list of cities, without visiting any city more than once. Interestingly, certain organisms like bees and ants are great at these kinds of problems, using vast numbers of individuals to search for a solution simultaneously.

In 2013, Google launched a Quantum Artificial Intelligence Lab with hardware from D-Wave. Google Glass represents the first practical use of their technology, with quantum technology helping its blink detector distinguish between intentional and involuntary blinks. Despite this progress, no one has yet created a general purpose quantum computer capable of tasks like factoring very large numbers, which is the basis of the RSA encryption that encrypts credit card numbers. You can only hope that before someone figures out quantum hacking, they will solve quantum cryptography as well.

What incentive would someone who created a quantum computer capable of breaking any code in the world have to share his or her discovery?

Qwerty Effect

Los Angeles Times
COIN RETURN
PULL AFTER COINS DROP
Los Angeles Times
Los Angeles Times

A Qwerty effect is a system that has become locked in as a result of subsequent technologies built upon it.

WHAT IF THE future of technology were shaped by decisions made long ago?

Next time you use your computer, take a closer look at your keyboard and ask yourself why the keys are arranged in that fashion. The placement of keys does not originate from an early personal computer, but is rather a relic of a typewriter layout designed by Christopher Sholes in the 1870s.

In the 19th century, typewriter designers had a challenge. They knew that a keyboard layout of ABCDE would result in more pairs of letters used, jamming the typewriter. That led to the rise of the Qwerty keyboard, which while slower, has persisted for over a hundred years, even though alternate layouts such as the DVORAK layout, have demonstrated increased typing speeds of as much as 70%.

Academics love to argue about whether Qwerty is a real example of the economic concept of path dependence. Paul David initially proposed that idea in 1985, and was critiqued by a wide range of researchers including Stan Liebowitz and Steve Margolis. Debate aside, Qwerty is an interesting example of the high costs involved if people needed to learn to type in a new way. It is also a great illustration of how network effects can lock in an entire industry once a technology becomes a standard.

Companies evaluating their technology should be on the hunt for Qwerty effects in their own architecture. Legacy systems, outdated business processes, even unexamined mindsets about how things should work, all contribute to dangerous lock ins that could prevent new technologies from being adopted in an optimal way.

It is hard to predict how many unexpected impacts an earlier technological decision may have. A recent study from the University of Chicago indicated that keyboard layouts may even have influenced how people name babies. The researchers argued that since the wider use of Qwerty keyboards on home PCs from the 1990s, more names have been picked from letters on the right hand side which are said to be more positive as it is the dominant side for most people.

What is an example of a self-imposed limitation in the way your business operates that may be no longer relevant?

Radical Transparency
Remote Work
Re-shoring
Retro Modding

Radical Transparency

Radical transparency is the philosophy of extreme information sharing inside an organization.

WHAT IF EVERYONE knew everything inside a company?

Some of the biggest sources of stress inside organizations are secrets. When a new boss is appointed, merger talks begin or layoffs are rumored, everyone goes into threat response mode. Personal survival is paramount, and people lose focus and stop creating valuable work.

Some companies are seeking to address this issue by making as much information available as is legally possible. The social media startup Buffer for example, publishes a report every month detailing the growth in its user base, revenue and total cash position. It publishes how much is paid to its workers and how it calculates those numbers. Strangely, rather than deterring employees, this approach has actually led to more job applications.

When I interviewed Brian Halligan, the CEO of Hubspot, he told me the story of how a few years into starting his business he had joined a group of CEOs who met regularly in Boston. At one particular meeting, the topic was culture. Brian had gone to the session thinking that the subject was a complete waste of time, until he sat next to the CEO of iRobot – who was not only brilliant but read him the riot act on culture. That same week, Hubspot surveyed their employees, and discovered that only half of them really liked working there. At that point he and his team decided to study the issue intensely. Their key assumption was that if the way people work and live had changed, the way they led and organized people also had to adapt.

Brian quickly realized that the real issue was transparency. Secrets created politics, especially in fast growing companies. He wanted his employees to trust him, and for him to be able to implicitly trust them too. They started sharing more information. They posted the notes from boardroom meetings on the wiki. They adopted a single policy of "use good judgment". Employees could submit questions, which would be handled at the company meeting. In Brian's words, they had their toughest employees monitor the founders at this session, to make sure that none of them got let off the hook.

How much time do the people in your business spend worrying about their own survival rather than the survival of your business?

Remote Work

Remote work is the performance of a job outside the usual environment of an office.

WHAT IF THE future of work were not in the workplace?

One of my favorite hotels in NYC is the Ace Hotel in Midtown. It's authentic, has fantastic coffee, and the lobby is a great place to hang out. There is only one problem. Almost everyone in the lobby is not actually staying in the hotel. They are all "working from home".

21st century work introduces a strange paradox. Mobile devices and cloud platforms give people the ability to work from anywhere, and increasingly, companies are encouraging employees to work from places other than the office. And yet many remote workers are electing not be at home at all, but rather spend their time in co-working spaces including coffee shops and hotel lobbies – places that curiously resemble the vibrancy of an office environment, just not their actual office. All of this begs the question – if remote work doesn't mean people want to be alone, what kind of spaces do they want to be in?

When Marissa Mayer, the new CEO of Yahoo, announced an end to the company's remote work program in February 2013 the news was met with shock. How could a technology company be against the principle of being virtually present? And yet, a few months later both Best Buy and HP announced similar programs. Mayer, as well as Meg Whitman, the CEO of HP, have both argued that although people may be productive alone, they are more collaborative when they are physically together.

Of course, not everyone believes that creativity is actually that important. Jason Fried, founder of Basecamp and a frequent protagonist in favor of virtual work, argues in his book Remote that most work is not coming up with the next big thing, "rather it's improving the thing you already thought of six months – or six years – ago. It's the work of work". In most cases, that can be better done alone.

As we design workspaces for the future, it should be taken as given that people can and will work from anywhere. The issue is not whether we should support the idea of remote work, but if we are going to have an office – how do we reimagine it in such a way that people will choose to actually spend time there.

What new skills would a manager of remote workers require that would not be necessary in a traditional office environment?

Re-shoring

Re-shoring is the act of bringing a business operation back to its country of origin.

WHAT IF IT made more sense to make things at home?

In 2010, 18 employees making iPhones and iPads at the Foxconn factory in China attempted suicide. With a workforce of 1.4 million, Foxconn is the largest private sector company in China and one of the world's largest employers. Conditions at the factory were bad, and unsurprisingly did not improve with the installation of suicide nets. In the end, with growing worker unrest and global scrutiny, management doubled pay at its factory complex in Shenzhen.

The troubles at Foxconn are emblematic of a broader shift in attitudes towards offshoring. Labor costs are rising in many emerging countries including China, which experienced a 10% increase in 2013. Companies are now taking a closer look at the total cost of manufacturing - which includes items such as transportation, reputation, intellectual property risks and the cost of carrying inventory.

I met Andrew Liveris, the CEO of Dow Chemical while I was speaking at the Global Competitiveness Forum in Saudi Arabia. Over breakfast, Andrew explained that the reason that their headquarters were still located in a small US town was because he believed in community. His factory workers, and their executives had their children in the same schools. But that was only part of the story. Dow Chemical has also commenced construction of one of the largest integrated chemical manufacturing complexes in the world – not in an emerging market but rather in Freeport, Texas. Community is important, but given the increasing supply of US shale gas, so are economics.

Falling energy prices have driven other re-shoring decisions in the US. Nucor Corp, which utilizes natural gas in its steelmaking process, has announced a new facility in Louisiana. Caterpillar is investing $120 million making excavator machines in Texas – work it was previously doing in Japan.

In the end, the decision to re-shore manufacturing will depend on the product in question. High value products, such as cell phones, can be airfreighted economically. But goods that have a low labor component and have to travel far by ship and be moved through road networks, will be prime targets for bringing back to home markets.

Aside from labor costs, what other factors should determine the geography of a global company's activities?

Retro Modding

HOUSE
TOYS AND BATTERY OPERATED PODUCTS
KAROL BAGH NEW DELHI-5

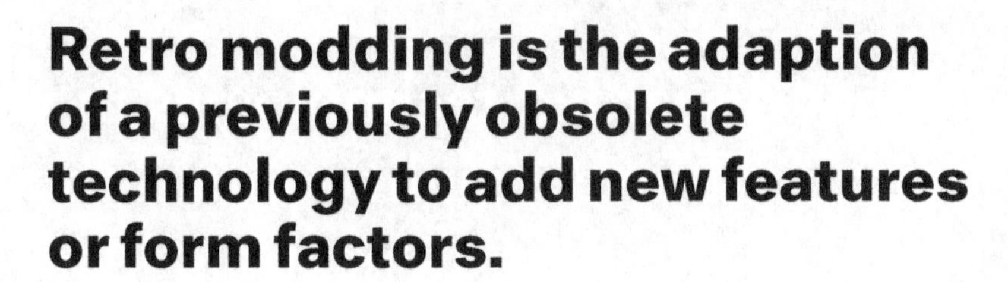

Retro modding is the adaption of a previously obsolete technology to add new features or form factors.

WHAT IF OBSOLETE technology could be brought back to life?

The retro modding scene in America has its own celebrity in the form of Ben Heck. Millions have watched his YouTube show where he hacks devices from phones to wheelchairs. Generally his targets for adaption are old game consoles. Heck deconstructs them and morphs them into new, miniaturized, bonsai-like configurations. Whether it be turning an old Super Nintendo into a handheld or reincarnating an Atari 2600 game system, Heck's adaptions are possible with the use of modern components, better screens, and smaller, less power-hungry elements. Heck calls his work "portabilizing".

One way to think of retro modders is as an example of what sociologist Henry Jenkins calls "participatory culture", where gamers cross the threshold of consuming content to re-mixing it. But as 3D printing, low-cost components, and online knowledge communities make modifying old technologies easier to accomplish - modding may become a powerful springboard to innovation.

Palmer Luckey was another teenager influenced by Heck. He set up a forum called Mod Retro, and in between repairing and selling old iPhones for money, started buying up and modifying old virtual reality gear from the 90s. In classic modding style, he took the equipment he bought apart to see if he could put it back together with more modern parts. He cannibalized components from his collection of head-mounted displays, and created prototypes with brand new screens. These projects were the basis of what would eventually become Oculus Rift, a company that sold to Facebook for $2 billion in 2014, just 18 months after raising $2.4 million on Kickstarter.

There are close parallels between modders and the Maker Movement. Both modders and Makers refuse to see technology as a hermetically sealed package, but rather as a book to be opened, studied, adapted and re-purposed. As in the case of virtual reality, companies may abandon a technology for economic or strategic reasons, but this doesn't mean that their prior investments can't be a platform for continued innovation by a community of hackers and tinkerers.

What discarded technologies in your bottom drawer might contain the seeds of the future's next big idea?

Self Driving Car
Shanzhai
Sharing Economy
Smart Tattoo
Social Graph
Space Commercialization
Syntetic Neocortext

Self Driving Car

10
20
30
40
50
60
70

A self-driving car is an autonomous vehicle that utilizes sensors and pattern recognition to navigate normal road conditions.

WHAT IF YOUR next car had no steering wheel?

The latest generation of luxury vehicles already includes connected services and auto-drive capabilities such as automatic cruise control and lane keeping. However the real question is not whether our cars will continue to help us drive, but whether in the future they will become entirely autonomous.

There are strong economic and social motivations for autonomous vehicles, with an estimated 34,000 people killed each year on the roads in the US. While it may seem counterintuitive to imagine that a robot car would be less dangerous, statistically that may prove precisely the case. It's notable that of the two accidents so far involving a Google self-driving vehicle, one was when it was being driven manually and the other was as a result of a rear end collision. Both cases were not the robot's fault.

Just as compelling a reason for change may be the benefit of having an entire network of connected vehicles, which collectively could allow better management of traffic flows, reduce parking demand in major cities, and in the case of automated delivery vans, take out one of the major contributing factors to congestion in metro areas.

The self-driving car group at Google was formed from the winning team of a DARPA contest to build an autonomous vehicle. Google cars work with a lidar (light radar) system, which enables the vehicle to generate detailed 3D maps of its environment. These, in combination with other high-resolution maps of the world, allow the car to produce the data models necessary for it to drive itself. Night or day, the car is able to consider its surroundings in over a gigabyte of data a second. Google's cars have safely driven more than 700,000 miles, but the next step is the design of vehicles that do not even feature a steering wheel or pedals.

The real potential for self-driving cars lies in "mobility on demand". For car manufacturers, that means switching from selling steel to potentially offering a subscription to transportation services. Then the consumer question is not what car to buy, but what is the right vehicle for this trip?

Should an autonomous vehicle have the power to allow its passenger to die in a crash, if it meant saving two other lives in an oncoming vehicle?

Shanzhai

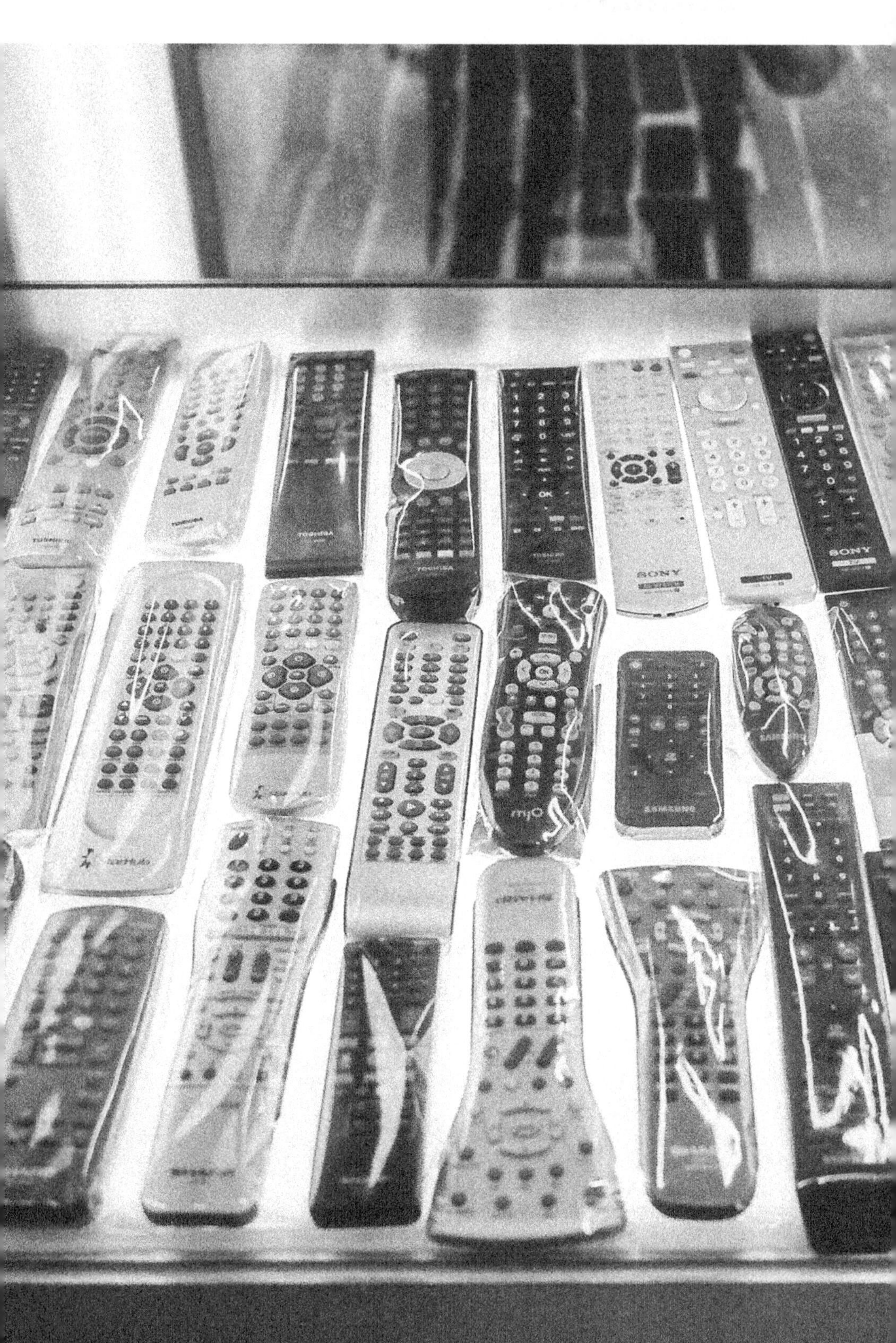
SONY
SONY

Shanzhai are low cost devices originating from the Shenzhen ecosystem.

WHAT IF A copy were better than its original?

Valerio Cometti, an Italian industrial designer once told me the story of how he was invited to China to help local manufacturers improve their design thinking. One by one, they brought in their products, placed them on the table in front of him, and to his astonishment, asked him what he would do to improve them. The Chinese approach to products, he realized, was very different to that of the West. Rather than taking the time to create a single beautiful tapware range, they would instead fill a catalog with thousands of designs from Baroque swans to spaceships.

You can see a similar logic at work in the surreal world of Shanzhai devices. Shanzhai, which literally translates as "mountain stronghold", is now used as a description for the wide array of devices originating from the Shenzhen ecosystem. Once a term used to suggest something cheap or inferior, Shanzhai has come to represent a certain Chinese cleverness and ingenuity.

Shanzhai devices are generally modeled on a famous original concept. Three months before the iPhone 5 was announced, a local manufacturer managed to not only steal the design, but also release its own version, the gooPhone i5. To add insult to injury, they immediately copyrighted and threated to sue Apple. Not all Shanzhai designs are direct copies. Many seek to "improve" on the original with the addition of TV tuners, dual-SIM card support, higher resolution cameras, solar chargers or telephoto lenses.

You might be tempted to dismiss Shanzhai as high tech counterfeiting, but this is to miss the point. Shenzhen is ground zero for a highly flexible and rapid manufacturing ecosystem. Shanzhai makers may have started out as copiers, but they are rapidly evolving into new types of customer-focused companies. Xiaomi is a case in point, being the first mainland brand that local teenagers actually want to own. In 2013, it sold 18.7 million smartphones; in the first half of 2014, 26.1 million. The phones are sold online using flash sales strategies, and priced near cost. They make money from software, accessories, and expanding margins as the cost of components falls over time. In other words, a very different kind of device company.

What is something that your customers would love to change about your product that a Shanzhai manufacturer would not hesitate to do?

Sharing Economy

51125

The sharing economy is based on the idea that underutilized personal assets should be rented out to other people.

WHAT IF YOU no longer owned anything?

One of the most extraordinary ideas that accompanied the birth of the Web, was that you might be able to trust a complete stranger. eBay was built on the belief of its founder Pierre Omidyar that "people are basically good". And yet what made eBay effective was that it didn't actually need that basic goodness in order to work. The most valuable thing for an eBay user was their reputation. User ratings were an effective, decentralized model of trust.

Today, similar models of trust have been used to provide access to a wide variety of goods and services. You can receive a lift from a stranger, or lend them your power tools, your car, or your boat. You might even allow them to stay in your house or eat your food. In the sharing economy, anything you own can be rented. Think of it as ownership unbundled from access.

Part of the boom in the sharing economy was the financial crisis – when suddenly people found themselves with assets they could no longer afford to own, or the need to find more cost effective ways to get things done. Still, getting users to behave in the right way required some new tricks.

My father was a retailer, and when I was a child he used to lead me around the department store he managed and do his best to impart life lessons to me. One day he showed me the mirrors that adorned all the walls. "Do you know why these are here?" He asked. "So people can see what they look like?" I replied. He shook his head and smiled. "People are less likely to steal things when they can see themselves doing it," he explained.

The social graph and the sense of being "virtually" visible has been one of the key innovations in the sharing economy. Linking your Facebook profile to your account provides a degree of accountability that was previously not possible among people who didn't actually know each other.

The growing popularity of sharing raises interesting questions about the future of ownership. Will products like cars and homes be designed with the expectation that their owners will also be entrepreneurs? Or will the next generation completely eschew the idea of high value assets in favor of a more nomadic life of on-demand access?

What would change if you knew the opportunity cost of all of your assets that you don't share?

Smart Tattoo

21-22
渋谷区
神宮前三丁目
21
Shibuya-ku
Jingumae
3-chome
21

Smart tattoos are sensor-embedded devices thin enough to be attached to your skin.

WHAT IF TECHNOLOGY were part of you?

When Tim Cook introduced the Apple Watch, many hoped it might do for wearable computers what the iPhone did for smartphones. And yet, great design aside, smart watches, bracelets and jewelry all fail to answer a more basic question: do people really want to carry yet another digital device?

That's why the most disruptive form of wearable computer may one day be those that are small and thin enough to be embedded on your body. Boston-based MC10, has been working on a range of "stretchable electronics" that can be applied internally to human organs, or externally to human skin or clothing. Their Biostamp device is thinner than a band-aid, while its sensors can monitor your personal vitals including temperature, movement and heart rate. The company is also developing a patch for tracking hydration and one that alerts you when it's time to reapply sunscreen.

The health monitoring capabilities of smart tattoos will make them essential for the management of conditions such as diabetes. Rather than the inconvenience of finger-prick blood glucose sampling, ultra thin sensors will read a patient's blood markers on a continuous basis, sending signals so that the user knows when to administer insulin.

Aside from health, smart tattoos may also offer a discrete way to control our devices. Motorola Mobility, owned by Google, has filed for a patent covering a tattoo that can intercept subtle voice commands, subvocal commands, and even subconscious thoughts. The device reads an auditory signal from the tattoo, placed over your vocal cords, and can use this to provide instructions to your other systems.

One of the major challenges of smart tattoos is power. Adding an external battery is not practical for daily use. One approach that may be useful for sports applications was proposed by Joseph Wang, a researcher from the University of California, San Diego. His solution was to simply use sweat. Sweat secretes lactate acid and electrons. By embedding enzymes that process lactate acid to a tattoo, it may be possible to produce 70 microwatts of energy per square centimeter of skin.

How might your use of technology change when you can forget it is there?

Space Commercalization

LIVE ON THE
EL CAPITAN
STAGE!
Jimmy Kimmel
For tickets call:
866-JIMMY TIX
abc

The commercialization of space is the shift of funding and development of the space industry from the public to the private sector.

WHAT IF OUTER space were the next big technology boom?

Since John F. Kennedy declared he would put man on the moon and return safely within a decade, the space industry has been driven by government spending and public contracts.

The commercial sector has long benefited from these investments. Spin-off technologies including memory foam, GPS, scratch-resistant lenses, aircraft anti-icing systems and even freeze drying all had their origins in the space program. In some areas, like satellite development, private companies chose to invest directly. Telstar 1, which was launched on top of a Thor-Delta rocket on July 10, 1962, was the first of many private satellites and on the day of launch relayed television pictures, telephone calls and fax images.

Recent changes to NASA's approach to its space program have shifted the focus of innovation to the commercial sector. The Commercial Orbital Transportation Services was developed to encourage private alternatives to the retired space shuttle fleet and to transfer cargo to the International Space Station. Both SpaceX and Orbital Sciences Corporation have been early innovators in this field. SpaceX's Falcon 9 rocket and Dragon cargo capsule made its first delivery in May 2012. Next up is the Commercial Crew program, which will focus on flight services and end the current reliance on Russia for rides into low orbit that cost nearly $70m per person.

While private companies have a strong incentive to innovate to bring down costs, ultimately there must also be larger rewards on the horizon to sustain further investment. Peter Diamandis, head of the X Prize Foundation has pointed to the potential of mining S-type asteroids, composed of iron, magnesium silicates, cobalt and platinum. An average half-kilometer S-type asteroid may be worth more than $20 trillion.

Other areas to watch include space tourism, as popularized by Richard Branson's Virgin Galactic, as well as hypersonic transport planes that travel in the upper atmosphere near the edge of space.

If the first space race put the US on an accelerated technology curve, what benefits might tomorrow's commercial space operators gain from their investments?

Synthetic Neocortext

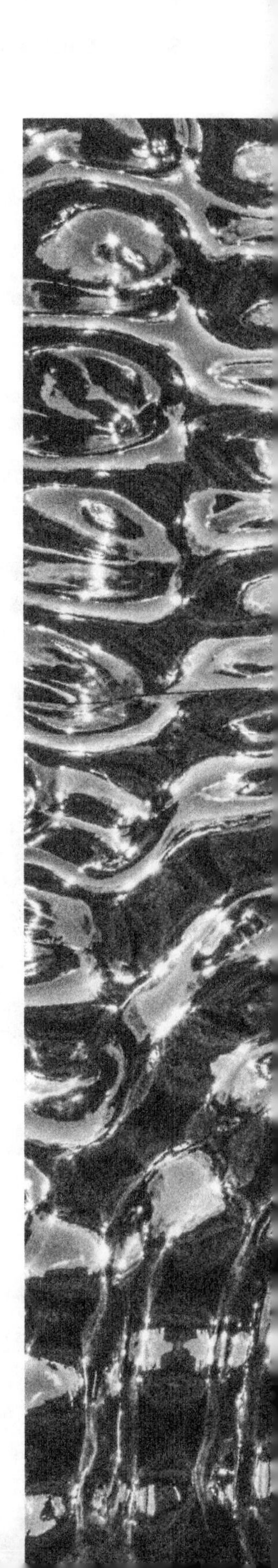

A synthetic neocortex is a computer-based simulation of the cognitive component of the human brain.

WHAT IF WE could reverse-engineer the human brain?

The neocortex is an extraordinary thing. It is what gave mammals their evolutionary edge over their instinct-bound competitors. And in humans, it represents about 80% of our brain. If you unwrapped a neocortex, it would fold out like a large table napkin. It is where our higher brain functions like sensory perception, conscious thought, reasoning and language originate.

Ray Kurzweil, a futurist and technology pioneer, believes that the neocortex is made up of about 300 million pattern recognition circuits that are responsible for much of our thinking. These circuits learn, recognize and implement patterns. All of these modules are arranged in hierarchies, which we start building the minute we are born. There are different levels of conceptual hierarchy, from recognizing a word, to recognizing a sentence, to remembering the name of a person walking into a room, or identifying whether something is funny or ironic.

Ultimately, we may be able to create a computer version of this design with sophisticated artificial intelligence algorithms that could exceed the hard limits of the human brain. The massive amounts of computing power required to achieve such a goal is part of the reason why Kurzweil joined Google as a director of engineering. There are other projects also aimed at creating an artificial brain, such as the Blue Brain Project, which is attempting to reverse-engineer a mammalian brain down to the molecular level.

Kurzeil believes that in the future we may be able to augment our thinking by creating gateways in our brains to pattern recognizers stored in the Cloud. In doing so, we would have access to vastly greater storage and computational power. Future search engines could listen to our conversations and monitor our actions, and suggest things we might need without us even asking.

According to Kurzweil, the quantitative leap in processing power from primates to humans is what gave rise to language, art, music and science. So given another leap in computation, what might we come up with next?

If everyone had access to the same set of computer augmentations, would this lead to more creativity or standardization in thinking?

Talent Density

Talent Density

田園書屋
高電
收
價話
園書屋
尚书房

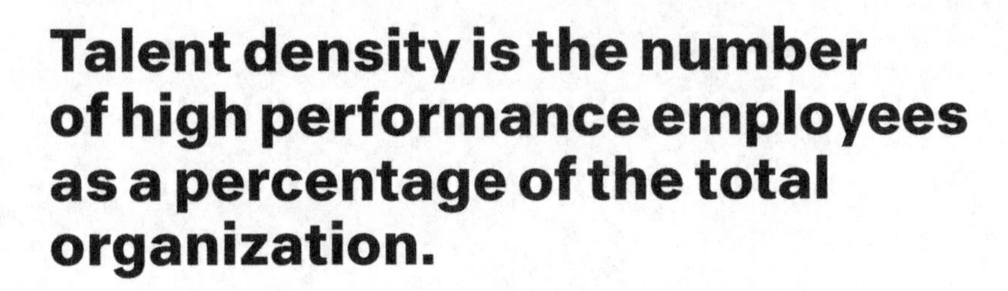

Talent density is the number of high performance employees as a percentage of the total organization.

WHAT IF YOU only needed a small number of people to do extraordinary things?

When Facebook acquired the mobile messenger service WhatsApp for $19 billion in 2014, it had grown to 420 million monthly users in just four years. More impressive was the number of employees – just 55. Instagram, bought for $1 billion, had 13. YouTube, bought by Google in 2006 had 65. Compared to telco or media firms, the differences in headcount are astonishing, but also raise a bigger question – what is the right way to grow a company?

In 2009, under pressure for a number of bold decisions, the CEO of Netflix, Reed Hastings, decided to make public an internal slide deck about how he hired, fired and rewarded employees. It became known as the "Netflix Culture Deck", and has been viewed over 8 million times on Slideshare. It was nominated by Facebook COO Sheryl Sandberg as one of the most important documents to come out of Silicon Valley.

In the deck, Hastings challenges the idea of "growing up". There is a conventional view that growing firms must add significant processes and procedures to deal with increasing complexity. Unfortunately more process results in a decline in talent density, as the number of high performing employees starts to fall with total employment growth.

At first, that's not a problem. Process-driven companies are efficient, have strong market share, and make few mistakes. The mavericks have left the business by this stage, but no one notices until the market shifts due to new technology, competitors or business models. At that point, the company is unable to adapt because their employees are only good at following process, and the "company grinds painfully into irrelevance". The solution, according to Hastings, is to increase talent density faster than business complexity.

Knowing how many people are needed to get something done is not an easy problem to solve. One of my former employees used to give two people the same job like dogs in a pit, to see which one would survive. Startup companies have an incentive to get more leverage from their human capital. If there is a smarter way to do something, they no choice but to find it. And perhaps that is the point. What the small have to do to survive is sobering advice for those companies that are big enough to have forgotten.

Which of your competitors use new technology or business models requiring significantly less people?

Ultra Wealth
Urban Scaling
User Agreement

Ultra Wealth

32
VADO
9163 FRP

The ultra wealthy are individuals with multiple millions in investible assets managed at institutional scale.

WHAT IF THE future of wealth were its concentration in the hands of a few?

Walk through the luxury districts of London, New York and Singapore and you will see some of the most expensive properties on the planet – brand new, bought in cash, and completely empty. Capital is now truly global, and wealthy individuals from China, Russia and Brazil have started buying high-end property, both as a status symbol and a safe haven from the uncertainties of their home markets. According to Oxfam, the 85 richest people in the world are as wealthy as the poorest half.

While globalization and export growth has certainly helped pull some of the poorest people in emerging markets out of poverty, it has also helped create a new global class of the ultra wealthy. And they are getting even richer. The latest Global Wealth Report published by the Boston Consulting Group in June 2014 indicated that the liquid wealth of this group had increased by 20% in 2013.

Many of the new rich come from emerging markets. According to McKinsey, millionaires in Asia (outside Japan) will create $7 trillion in net new wealth by 2016. Even more interesting is that in China about 80% of millionaires are under 45 years old, compared with just 30% in the US and 19% in Japan.

As wealth concentrates, controversy and civil unrest will grow. The Occupy Wall Street campaign, which drew attention to the disproportionate influence of the 1%, was just the beginning. French economist, Thomas Piketty, roused further debate with his book Capital in the Twenty-First Century, in which he argued that a principle reason for mass inequality is that capital brings higher returns than wages.

While governments may seek to address the issue through taxation, the truth is that the ultra wealthy are more like institutions than individuals. They operate in the cracks between nation states, and with an army of financial and legal advisors can re-structure reality to their will. As a small but powerful group they will challenge everything from the pricing of metropolitan real estate to competition for entrance to the best educational facilities.

What traditional assumptions about wealthy consumers are no longer valid when your target market is young, rich and Chinese?

Urban Scaling

Urban scaling is a theory of how modern cities behave as they grow.

WHAT IF THE destiny of a city could be predicted with mathematics?

Human beings and cities have always had an unusual relationship. Our bodies have long been a reference point for both design and interaction. We measure length in feet, and pour scotch in fingers. Cities that seem vibrant and creative also often preserve that sense of human scale – they are pedestrian friendly, dense, and visually dynamic.

Human scale is a seductive idea, especially if you have the opportunity to plan an entire city from scratch as Le Corbusier did when he was commissioned in the 1950s to design the capital of Punjab. "The city of Chandigarh is planned to human scale," he said, speaking of his work. "It puts us in touch with the infinite cosmos and nature. It provides us with places and buildings for all human activities by which the citizens can live a full and harmonious life. Here the radiance of nature and heart are within our reach."

With a sales pitch like that, I couldn't resist a chance to see Chandigarh for myself. As I toured the city, I could see the impact of population growth. The signs of strain were everywhere. The city was curiously planned in the shape of a human body, and ironically, it was those very humans that eventually corrupted its vision. In the main government buildings, exposed wires snaked through the ceilings like creeping vines; priceless original furniture was stacked like firewood in the corners; offices overflowed with paper folders and ancient tea kettles. It was like a paean to a misplaced utopia, and proof that even the most beautifully laid plans rarely survive contact with the future.

So, if you can't reliably plan a city, can you at least calculate its trajectory? Geoffrey West, a theoretical physicist, spent two years studying the underlying data of big cities. He discovered that the characteristics of cities could be reduced to equations. For example, if you knew the population of a particular area, you could accurately estimate its average income and the dimensions of its sewer system. As cities grow, they obey certain rules. Scale matters. Urban, social and economic phenomena become more intensive with city size. And most importantly, as West and his colleague Luis Bettencourt have argued, cities are social reactors. The more people who live in one place, the easier it is to facilitate human interactions, the exchange of ideas and creative collaborations.

If you were designing an office of the future, what would you do to ensure its social reactivity?

User Agreement

A user agreement is a set of legal principles that govern the usage of a web platform or software application.

WHAT IF YOU gave away your rights and freedoms to play a free mobile game?

On April 1 2010, the now defunct retailer Gamestation added a clause to its order conditions stating that the customer agreed to irrevocably give their soul to the company. It may have been an April Fool's joke, but the point was serious. 7500 users agreed to the contract, even though there was a checkbox to exempt out of the "immortal soul" clause. It turned out that 88% of users did not read their agreements.

One of the consequences of software taking over so much of our daily lives is that we are now subject to the complexity of legal conditions that digital companies use to protect themselves. Sometimes called End User License Agreements, or "click-through" agreements, they are the fast-scrolling terms and conditions that load up after you buy a mobile app, or when you sign up to a social networking service. They are long, complex, and as Gamestation's immortal soul prank proved, largely unread.

User agreements are more prohibitive than you might expect. Terms frequently include the provision that your content and photos may be used for commercial purposes. Other contracts, notably for Apple's iTunes, require you to agree to every change in future versions of the software, with your consent implied should you simply listen to your own music library. In one example, a Web application had a dense 5700-word contract containing a clause that granted the software company the power to use the customer's spare computing power to mine Bitcoins.

It is reasonable to expect that software makers will seek to limit their liability and ensure that their applications are used for proper purposes. The problem arises when a user does not agree with some of the terms that require consent. There is no scope to join Facebook on a limited or negotiated basis. You either accept all the conditions, as they may change with time, or not. As consumers become aware of the value of their data and the threats to their privacy, we may in the future see greater push back on the aggressive use of user agreements to force compliance. Then again, like all negotiations – you only have as much power as your next best alternative.

How exposed is your business when your employees use consumer online applications while at work?

Virtual Reality

Virtual Reality

TOILET
好評受付中！
フレッツ・テレビ
ブロードバンド受付カ

Virtual reality is a set of immersive technologies able to create the impression of being inside a digitally constructed environment.

WHAT IF A simulation were indistinguishable from reality?

There was a point in the 90s when the arrival of true virtual reality seemed inevitable. There were frequent references in movies and popular fiction, as well as a number of military and consumer virtual reality (VR) companies in operation. But the boom was short lived. Blocky graphics, expensive rigs and nausea-inducing displays eventually led the industry to a technological dead end.

That is, until a 21-year-old named Palmer Luckey managed to reboot the entire category with Oculus Rift, a VR headset born of cannibalized parts from obsolete systems as well as state of the art components. Now there is a new and fast growing ecosystem of VR startups that make everything from motion sensing suits to treadmills that allow you to walk through virtual worlds.

One of the reasons that VR was able to re-emerge anew is that this time the technologies are being built on top of existing applications, mobile technologies and online platforms. In the case of gaming, a lot of the content was already in place, and all that was missing were better interfaces.

The next step for VR will be not just representing reality, but translating human motion and interaction into the digital realm. Philip Rosedale, founder of the virtual world Second Life, recently launched a new project called High Fidelity, the focus of which is simulating realistic human expressions with avatars capable of tracking your movements and gestures. But that is just the beginning. Using sophisticated motion capture technologies, you will eventually be able to project your head movements and facial expressions onto an avatar. Or with a full-body harness, you could literally step into a digitally created world.

Rosedale's other big idea is to leverage the distributed processing power of users to power his simulation. In exchange for virtual money, virtual citizens can assign their computer's unused processing power to help render High Fidelity's world in detail.

What might be the consequences for the real world if people spent most of their time in a virtual one?

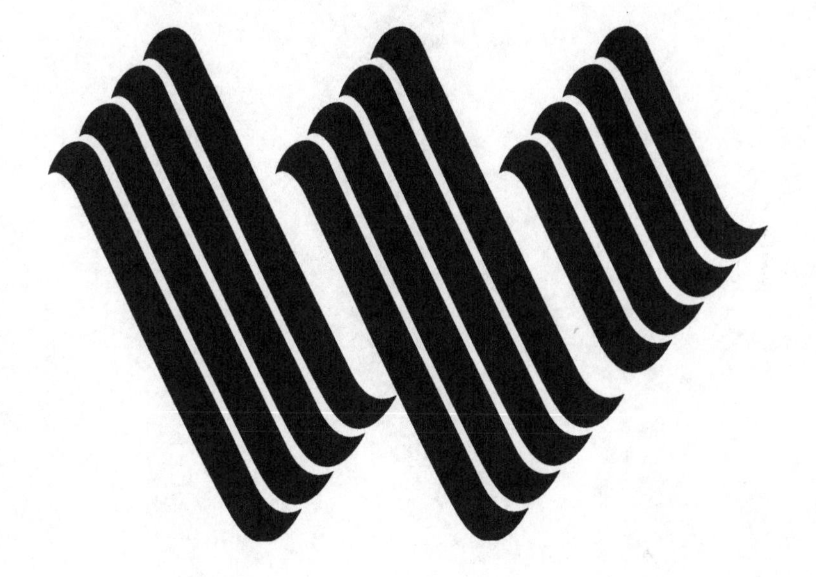

Webscale
Workforce Analytics

Webscale

Pizza Hut

Webscale is an approach to building IT infrastructure that can handle exponential growth.

WHAT IF YOU let Google build your infrastructure?

When Instagram launched in October 2010, they signed up 25,000 users on their first day. The founders had to call in all of their Stanford University friends to help, and it took the entire group working around the clock to keep the servers from crashing. They transferred their site to Amazon, and a month later, it could handle a million simultaneous users without a problem.

In order to achieve their success, the major "Cloud Builders" like Google, Amazon, Netflix and Facebook, have not only created innovative consumer products, but taken radically different approaches to IT infrastructure as well.

Cloud Builders tend to build their infrastructure over short periods of time, and from the ground up. They use open source technologies, agile development processes and, because of their need for massive scale and cost control, actually design a lot of their hardware themselves. These units are stripped of extraneous features and provisioned virtually

Traditional companies, on the other hand, build their infrastructure over long periods of time, integrating legacy systems and accommodating complex support contracts, licenses and strict guidelines on the use of IT services. The complexity of these arrangements not only makes it expensive to add new capacity, but may also mean that the new equipment will not be provisioned in time to meet sudden spikes in demand.

Unfortunately, sudden shifts in user behavior are part and parcel of today's network world, which Reed's Law asserts "can scale exponentially with the size of the network." As big companies increasingly compete with innovative software and customer applications, they will face a choice. Can they change their culture and mindset to embrace a Webscale IT model, or will they have to rely on outsourcing large components of their technology operations to those platforms that can?

What would cause you the most grief: your next app going unnoticed or becoming an overnight viral success?

Workforce Analytics

Workforce analytics is the use of data to study the performance and capabilities of employees in the workforce.

WHAT IF YOU could use data to predict employee behavior?

When I finished law school, I almost took a job as a consultant at McKinsey. Fortunately for them, the arrival of the Web threw a spanner in the works, and I ended up running a startup instead. But I will always remember my last interview with the company. After dozens of probing sessions, mind games and analytical puzzles, I found myself across the table from the managing partner of the business. He lent forward and after a pause asked intently, "What do you think that we think your weaknesses are?"

All leaders have their own secret combination of intuition, checklists and questions that they hope might unlock the secret to a great hire. What is interesting however, is how many companies are now trying to back up those gut feelings with the reassurance of numbers.

Talent analytics uses algorithms and data science to predict factors critical to performance, in order to build a model for future hires. Consider some of the loose assumptions companies have made for years. They hire people with good grades, but what is the point of hiring academic sales people if the best performers are those with previous experience selling expensive items with long sales cycles? Companies analyze their past data, but what is the point of knowing how many people have left your company when you can identify behaviors that might suggest someone is an immediate "flight risk"?

Google has made considerable investments in what it calls its "People Operations". As a company, it deploys the same kind of rigorous testing and statistical analysis to the problem of human performance as you might expect to find in a research lab. One of Google's algorithms for example, enables the prediction of candidates with the highest probability of succeeding after they are hired. Google has also calculated that there is very little incremental insight added beyond four interviews, although it is essential that hiring decisions are made by a group in order to prevent individual managers hiring for short term needs. The People Operations team also study more abstract questions, like how to encourage people to be more collaborative, or whether asking staff to leave their phones in the office might improve work/life balance.

If you don't know why your people leave, can you determine how much value your former employees might create for your competitors?

XAAS

XAAS

XAAS, or "everything as a service" refers to the delivery of a wide range of software applications and services over the Internet.

WHAT IF ALL of your business needs were available as on-demand services?

The Cloud has changed the way companies buy and use software. Lower costs, pay-as-you-go billing, limited capital investment and faster launch times – all are strong arguments for the software-as-a-service business model. Essentially, what makes the Cloud really compelling is that it allows business leaders to consume IT, rather than just invest in it.

On-demand software has expanded rapidly to include services as broadly encompassing as infrastructure-as-a-service, platform-as-a-service, storage-as-a-service, desktop-as-a-service, disaster-recovery-as-a-service, and even business-processes-as-a-service. In my view, that final idea is the most interesting.

For 21st century companies, the ability to consume business processes via the Cloud will become increasingly important. Companies have experimented with both automating and outsourcing business processes for decades – whether it be accounts, payroll or credit checks. In the future however, even complex processes such as shipping and distribution will be packaged in a way that integrates external partners, relevant data and analysis, and made available as a service through a Web browser or mobile device.

The shift to consuming business processes rather than building and managing them, will allow organizations to focus on more compelling priorities. If you can subscribe to payroll and benefits transactions, then HR teams can spend their time identifying top performers and developing leaders. If IT can subscribe to customer care and security management, then they can shift their attention to aligning more closely with the business.

XAAS is more than just a change in pricing and technology delivery. It is a future model for business that allows for greater mobility of employees as well as the ability to scale up quickly without additional infrastructure costs.

If you were designing your company today from the ground up, what business services would you subscribe to rather than build?

Yottabyte

Yottabyte

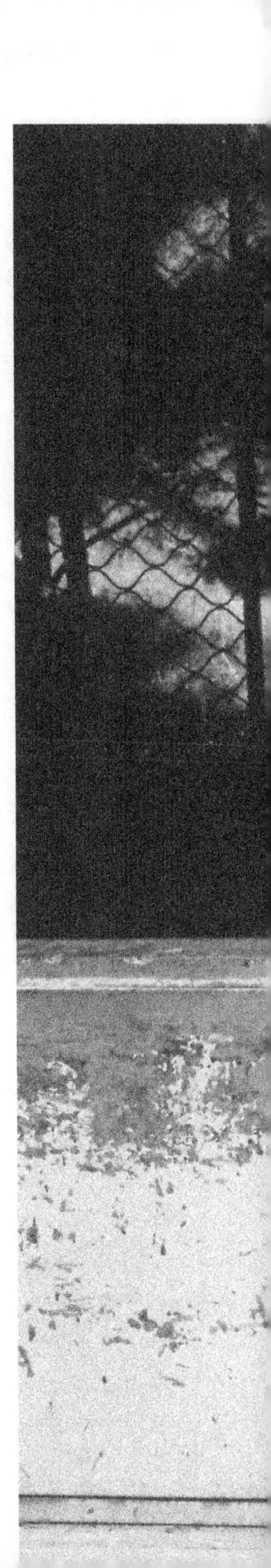

All kinds of
memory sticks
battery film
camera postcar
BEIJING
The Ming Tombs
THE FORBIDDEN CITY

A yottabyte is a billion petabytes.

WHAT IF NOTHING ever needed to be deleted?

A yottabyte is about as large a data metric as we can currently imagine. To put that number in perspective, it is a billion petabytes, a trillion terabytes or in more evocative terms, 1,000,000,000,000,000,000,000,000 bytes.

At any point in a society's technological evolution, large data metrics appear ridiculous. The idea that we could have ever filled a terabyte of information was inconceivable in the 1960s when mainframes were designed for megabytes. And yet now, we carry terabyte drives around in our bags as spare storage for our music collections.

The digitization of both our lifestyles and businesses has contributed to an explosion in data storage. According to IBM, every day, we create 2.5 quintillion bytes of data. In fact, 90% of the data in the world today has been created in the last two years alone. Facebook's data warehouse stores over 300 petabytes of data, with 600 terabytes added daily. The CERN Large Hadron Collider generates 1 petabyte per second, while sensors from a Boeing jet engine create 20 terabytes of data every hour.

In the near future, new health and genomics research will require even greater levels of storage. The Brain Research through Advancing Innovative Neurotechnologies (BRAIN) is an initiative to "map" the human brain. Francis Collins, director of the National Institutes of Health, said the project could eventually entail yottabytes of data.

What all of this adds up to is the strange proposition that data might last forever. Even though our human lives are transient, all of our actions, transactions, communications and content may never be completely deleted. Think of it - the sum total of a life stored for all time on a database in the middle of the desert or on any icy landscape deep in the Nordics. How much would a single life take up on a yottabyte hard drive? A depressingly small amount.

If you stored a copy of everything you ever saw, did or said, what kind of application would you need to make that data useful?

Zero Email

Zero Email

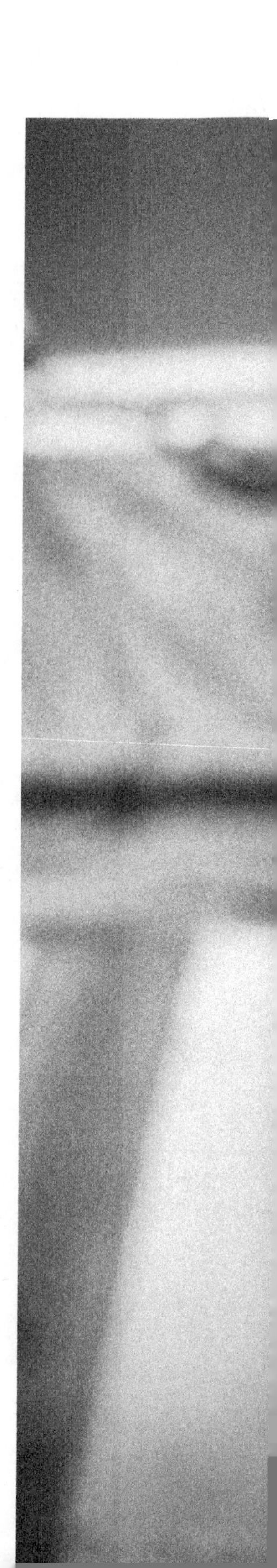

REDIAL

Zero email is the goal of eliminating the use of electronic mail communications in an organization.

WHAT IF NO one used email anymore?

The idea of a world without email is not so dangerous to some people, especially those below a certain age range. I remember asking a twelve year old whether he used email. He laughed and said only when he had to talk to his parents.

The first email was sent by Ray Tomlinson, a US programmer who implemented a messaging system in 1971 on ARPANET, the forerunner of the Internet. By all accounts the contents of that first email were not particularly interesting, but the system itself must have been because Larry Roberts, a director of DARPA, started using it for all his communications. Any researcher who needed his funding approval had to use email, and before long, it became a widely adopted tool.

Several decades later, email is an unavoidable part of daily work life, but also one that is fraught with problems. When the best people in an organization collaborate over email, those insights are forever lost or hidden to others. As for the worst employees? Well, email is a corporate politician's best friend.

Atos, the French software company, announced a program in 2011 to become a Zero Email organization. Their view was that the best way for their 76,000 employees to work together was by using next generation tools like enterprise networks, instant messaging and document sharing platforms - anything but email. Since starting the program, they have launched 7500 purpose build communities, with 30% of active users posting 10 collaborative notes per week. They estimate that $130 million in additional deal value has been unlocked as a result of faster responses to bids, the ability to source internal experts and information more quickly, and the reduction of their reaction time, in some cases, from 2 days to 45 minutes.

Increased productivity is only a small part of the idea of eliminating email. The other more important issue is one of work/life balance. When Google discovered that more than two-thirds of their employees were unable to stop thinking about work after leaving the office, the company ran a trial in Dublin called Google Goes Dark, in which staff were asked to leave their work devices at reception when they went home for the day.

If email were banned at your business tomorrow, what impact would it have on your company culture and productivity?

Acknowledgements

THE PRODUCTION OF this book was a dangerous idea in itself. From start to finish, A to Z - writing, editing, design and production – all done within a month. This would have been an impossible goal without the brilliant and talented team that agreed to help me.

Bold, based in Stockholm, conceived the layout and the visual feel of the dictionary. They are one of the best design agencies I have ever worked with. Gemma O'Brien, a typographic rock star and acclaimed illustrator, created the entirely new alphabet featured in this project. Lucy Howard-Taylor, a famous author in her own right, did a wonderful job untangling my convoluted thinking by editing my manuscript.

Most importantly I would like to thank the one person in my life who inspires me daily – my beautiful wife, Burcu. When we met, I signed her a copy of my first book and wrote that I never wanted to imagine a future without her in it. As a futurist, I've been wrong about more things than I care to admit – but that is the one prediction that is truer today than it ever was.

CPSIA information can be obtained at www.ICGtesting.com
Printed in the USA
LVOW02s2306251114

415664LV00022B/195/P